Engineering for the Control of Manufacturing

PRENTICE-HALL INTERNATIONAL SERIES
IN INDUSTRIAL AND SYSTEMS ENGINEERING

W. J. Fabrycky and J. H. Mize, Editors

ALEXANDER *The Practice and Management of Industrial Ergonomics*
AMOS AND SARCHET *Management for Engineers*
BANKS AND CARSON *Discrete-Event System Simulation*
BEIGHTLER, PHILLIPS, AND WILDE *Foundations of Optimization, 2/E*
BLANCHARD *Logistics Engineering and Management, 2/E*
BLANCHARD AND FABRYCKY *Systems Engineering and Analysis*
BROWN *Systems Analysis and Design for Safety*
BUSSEY *The Economic Analysis of Industrial Projects*
CHANG AND WYSK *An Introduction to Automated Process Planning Systems*
ELSAYED AND BOUCHER *Analysis and Control of Production Systems*
FABRYCKY, GHARE, AND TORGERSEN *Applied Operations Research
   and Management Science*
FRANCES AND WHITE *Facility Layout and Location: An Analytical Approach*
GOTTFRIED AND WEISMAN *Introduction to Optimization Theory*
HAMMER *Occupational Safety Management and Engineering, 3/E*
HAMMER *Product Safety Management and Engineering*
HUTCHINSON *An Integrated Approach to Logistics Management*
IGNIZIO *Linear Programming in Single and Multiobjective Systems*
MIZE, WHITE, AND BROOKS *Operations Planning and Control*
MUNDEL *Improving Productivity and Effectiveness*
MUNDEL *Motion and Time Study: Improving Productivity, 6/E*
OSTWALD *Cost Estimating, 2/E*
PHILLIPS AND GARCIA-DIAZ *Fundamentals of Network Analysis*
PLOSSL *Engineering for the Control of Manufacturing*
SANDQUIST *Introduction to System Science*
SMALLEY *Hospital Management Engineering*
THUESEN AND FABRYCKY *Engineering Economy, 6/E*
TURNER, MIZE, AND CASE *Introduction to Industrial and Systems Engineering, 2/E*
WHITEHOUSE *Systems Analysis and Design Using Network Techniques*

ENGINEERING FOR THE CONTROL OF MANUFACTURING

KEITH R. PLOSSL

Prentice-Hall, Inc., Englewood Cliffs, New Jersey 07632

Library of Congress Cataloging-in-Publication Data

Plossl, Keith R. (date)
  Engineering for the control of manufacturing.

  Bibliography: p.
  Includes index.
  1. Industrial engineering. 2. Engineering design.
3. Production engineering. I. Title.
T56.P53 1987   658.5   86-9485
ISBN 0-13-277773-8

Editorial/production supervision and
  interior design: Gloria Jordan
Cover design: Lundgren Graphics, Ltd.
Manufacturing buyer: Rhett Conklin

©1987 by Prentice-Hall, Inc.
A division of Simon & Schuster
Englewood Cliffs, New Jersey 07632

Printed in the United States of America

10 9 8 7 6 5 4 3 2 1

ISBN 0-13-277773-8 025

Prentice-Hall International (UK) Limited, *London*
Prentice-Hall of Australia Pty. Limited, *Sydney*
Prentice-Hall Canada Inc., *Toronto*
Prentice-Hall Hispanoamericana, S.A., *Mexico*
Prentice-Hall of India Private Limited, *New Delhi*
Prentice-Hall of Japan, Inc., *Tokyo*
Prentice-Hall of Southeast Asia Pte. Ltd., *Singapore*
Editora Prentice-Hall do Brasil, Ltda., *Rio de Janeiro*

To my wife and my parents,
for their support, encouragement, and help.

CONTENTS

3 DESIGN ENGINEERING   29

4 COMPUTER SYSTEMS   42

5 COMPUTER PROGRAMMING   60

12   PLANT ENGINEERING   179

13   FORECASTING AND MACHINE UPTIME   193

FOREWORD

In the urgently needed effort to revitalize manufacturing in the United States, there is no more important factor than the role of engineers. Those in product design, industrial, manufacturing, process, tooling, plant, maintenance, and others working in manufacturing hold several keys to unleashing the enormous potential of American industry to recapture its former position of leadership in worldwide competition.

This book is unique. It is the first to define clearly the new roles of engineers in redirecting manufacturing to much higher productivity, for better utilization of resources, and greatly improved customer service. It shows also the benefits engineers can realize through understanding and using modern business planning and control systems. It demonstrates with specific examples the "no lose" situation for all functions in a business.

Written by an engineer for engineers, this book distills Keith Plossl's two decades of experience in industry as the role of the engineer has evolved from that of a traditional technical specialist into that of a teammate required to implement Just-in-Time strategies. Challenging conventional wisdom as it does, this book provides engineers and managers in industry with much food for thought as well as suggested actions.

Engineers have been accused of seeking psychological sanctuary in scientific scholarship. This book will shock them into the real world.

George W. Plossl, P.E., CPIM*

PREFACE

This book was written for engineers, engineering managers, materials managers, and college students who are or will be working in manufacturing. It provides a thorough understanding of integrated engineering in a manufacturing company. The viability of such integration has been proven by successes both in the United States and abroad. Effective computerized system implementation in the manufacturing environment requires discipline and knowledge of how to integrate activities. What to do to achieve integration is the subject of this book. Many books have covered the specifics of manufacturing and engineering techniques, but they were written as if these techniques were independent of each other. This book attempts to show the interrelationships of manufacturing and engineering to the whole manufacturing environment. As George Plossl often urged, we must see "the big picture." Integration is neither easy nor fast. There are no magic techniques that produce instant results. It is the application and integration of time-honored techniques that improves manufacturing performance.

This book describes the fundamentals of the modern manufacturing control system, the engineering support required, and the use made of the system's tools. It covers these areas from a practical standpoint rather than a theoretical one, but theory supported by practical application and experience has not been ignored. The bibliography contains references that go back more than thirty years. When applied in an integrated way these techniques lead to the manufacturing productivity improvements companies seek today. We don't need exotic techniques; we need the fundamentals enhanced by modern computer systems. Complex computerized systems operated independently create manufacturing bankruptcy with blinding speed. Sophisticated computer systems that are used to integrate engineering, manufacturing, and materials control are clearly the keys to worldwide competitive manufacturing excellence. Engineers don't need another book on techniques; they need one that welds proven techniques into a coordinated, successful manufacturing strategy. This book addresses that integrated strategy.

Keith R. Plossl

1

ENGINEERING

IN A MANUFACTURING COMPANY

ENGINEERING IN THE MANUFACTURING ENVIRONMENT

Engineering is key within the manufacturing environment to product design, product manufacturing flow, and the ability of a company to produce its goods reliably. Engineering design really determines the appearance, function, cost of production, and the ability to plan and control manufacturing operations. Engineering helps determine—by product design, plant layout, standards, methods, tool design, and machine maintenance—the required levels of raw material, work-in-process, and finished goods inventory. However, one of the greatest challenges today is making engineers aware that they have this influence and overcoming their isolation from the manufacturing process.

The potential benefits of integrating engineering into manufacturing are enormous, as is evident from a survey of the international competition. This book will show how coordinated engineering can improve plant capacity by over 30 percent, increase return-on-assets tenfold, improve engineering and plant productivity over 50 percent, and do all this with one half the energy consumption. Overseas competition is operating with double- and triple-digit inventory turns and with more goods produced per employee hour worked than is the case in most American companies. In fact, international competition is increasing both the number of product offerings in general and the number of high-technology products, an area traditionally dominated by the United States. There is no doubt that U.S. smokestack industries are receding in prominence. High-technology manufacturing is under increasing assault from overseas competition. We know how to control manufacturing in order

1

to improve productivity, and the United States has the resources in men, materials, and money to compete more effectively. Coordinating engineering with manufacturing is key to success in this effort.

The problems of the manufacturing environment are complex and interrelated. The theoretical approach of optimizing individual departments and functions to produce the most efficient total operation has been disproved many times in practice. Another fallacy, now recognized as such, is that the relationship between quality and cost is directly proportional; in fact, it really costs less to make a high quality product, not more. If a company is to produce at optimal efficiency with the highest customer service and minimum inventory, manufacturing must operate like a well-oiled automated machine. Scrap will grind that machine to a stop.

The goal for the manufacturing company is clear: integration of functions and an understanding of the fundamental relationships between groups to produce the highest possible productivity of the whole. Companies that succeed in integrating engineering into their systems will realize great gains in productivity, significant reduction of inventory, and maximization of customer service; those who fail to integrate may not survive. Chapter 15 will present the theory and practice of production by the Just-In-Time approach. This stockless production system would make production approach the single machine concept and could make custom-designed equipment as efficient to make as mass-produced sameness is today.

COMPUTER ASSISTED ENGINEERING (CAE)

No discussion of engineering today is complete without some computer system and programming topics. Engineering has progressed from the slide rule through the programmable calculator to the computer. Increasing productivity in engineering today is seen as synonymous with computerized engineering. A word of caution: Although computers are worthwhile tools, we have seen them improve engineering productivity but reduce factory output at the same time. The integration of computer technology with engineering, materials control, and manufacturing is needed to effect productivity improvements in all activities.

Chapter 4, "Computer Systems," covers computer systems used in manufacturing and engineering. It will concentrate on Computer Aided Design (CAD) and the types of hardware available today for this process. The characteristics of such hardware are important for achieving integration between engineering and manufacturing.

Chapter 5, "Computer Programming," covers computer languages. An understanding of the history and development of some common languages is helpful in appreciating the difficulties in achieving integration of systems. The details of different programming languages are omitted; instead the chapter concentrates on the common characteristics of languages. The U.S. government is directing development of a common universal language called ADA, but it will take many years before this language can be utilized. In the meantime, engineering, manufacturing, and materials control must begin to share data among their computer systems. The language barrier is a significant but not insurmountable problem. This chapter provides the foundation needed to achieve solutions.

The CAD/CAM system and the CAE/Materials Control interfaces are so important that Chapter 6, "Computer System Interfaces," has been devoted to them. The use of networking to achieve integration and application of distributed processing are some keys to better flow of information. The day of the ever-increasing central mainframe computer is past. This does not mean that mainframe computers will disappear but entire organizations will not be dependent on single processors for all tasks. With the development of personal and mini-computers, providing engineering, accounting, order entry, purchasing, planning, warehousing, and sales each with their own computers and programs has become economically feasible. This technology has developed so rapidly, however, that most managements have not learned how to coordinate, organize, and manage data scattered all over the plant. Database management and system integration are essential for distributed systems.

PROJECT PLANNING AND CONTROL

Project planning and control is a common requirement of all three engineering departments. *Design engineering* has new product development projects. *Manufacturing engineering* has layout, new process equipment, and tooling projects. *Plant engineering* has facilities and maintenance projects. All three must have some scheduling system for those projects which permits management to monitor progress and allocate resources in a timely fashion.

Project planning and control differs little from materials planning and control in many respects. Both have a finite result in a finished product or project, a planned sequence of operations or tasks that are required to produce the result, and a limited supply of resources or materials to use. Project scheduling has used Gantt charts, Project Evaluation and Review Technique (PERT), Critical Path Method (CPM), and various other techniques.

An important development in project planning was the recognition that Materials Requirements Planning (MRP) logic can be applied. This logic permits both planning a single project and planning all the projects within an engineering department. MRP logic can be applied to scheduling *both* human and material resources, and it has the advantage of *not* assuming that resources will be available when needed. The Project Resource Requirements

Planning (PRRP) technique uses the material control system software and only requires that engineers understand MRP logic. This technique is discussed in Chapter 9, "Project Planning and Control."

THE HISTORY OF ENGINEERING IN MANUFACTURING

The dictionary defines engineering as "the planning, designing, construction, or management of machinery, roads, bridges, or buildings." In small companies, engineering may be represented by a single person who may or may not have any formal engineering education but who provides the company with expertise in design, equipment, plant layout, and maintenance. Companies are usually founded by a small group of people with an idea for a product they wish to bring to market. The group functions as a team incorporating their skills of product design, marketing, production, and sales. The pressures of competition in the marketplace and rapidly expanding technology result in the need to increase product complexity and sophistication. These pressures lead to the need for additional technical knowledge which can no longer be supplied by one engineer.

Historically, engineering departments have expanded through functional specialization. The long-term goal has been to divide engineering into the logical functional specialties of design, process, and facilities. As the small company evolves into a large company, this specialization becomes more pronounced and expertise is applied to ever narrower and increasingly technological fields. In large companies or companies with very complex products, the engineering department usually consists of very highly trained specialists working in distinct, functional groups. This functional separation provides experts, accountability, and convenient compartmentalization for control of activities; however, it often leads to the belief that engineering is a creative art so steeped in technology that manufacturing operations are of little interest or consequence. This aloof attitude develops expertise in technology but does little for manufacturable products.

For purposes of this book we will divide engineering into three functional groups: design, manufacturing, and plant. Design engineering is the group responsible for the physical design of the product. Manufacturing engineering is the group responsible for the process, the layout of production equipment, and the work methods and standards for managing the process. The plant group is responsible for the facilities and maintenance of the physical plant and machinery.

THE DESIGN ENGINEERING DEPARTMENT

Design engineering is viewed as having six fundamental roles in manufacturing:

1. Inventing new products
2. Improving present products
3. Providing documentation as to
   part numbers
   drawings
   specifications
   tolerances
   bills of material
4. Supporting marketing
5. Supporting manufacturing
6. Supporting material control

The invention of new products is a key to a company's future growth. This area is so important that many companies do not permit any outside distractions to enter their world. Thus, although the same people who develop new products may also improve present products, it is more often the case that there are two groups of engineers and technicians for these functions. The redesign function is a training ground for new design engineers in many companies.

The design engineering department is responsible for those activities conducted prior to actual manufacturing operations. These consist of research and development, product and machine design, and product redesign functions. These are highly creative activities, and the role played by this group in the manufacturing environment is greatly influenced by management's expectations. Frequently, management permits this group to function in an almost "ivory tower world" to optimize their creative talents and spur rapid design development. In fact, this development seems a direct outgrowth of the specialization needed to keep pace with expanding technology in the growing small company. Nevertheless, in most cases this suboptimization of activities makes for decreased overall plant productivity. The reasons for this will be evident later.

Providing documentation is the function design engineers dislike the most. Part numbering is not difficult for engineers although many are unaware of the implications of using long or significant-digit part numbers. Engineering drawings are necessary to define the product along with part specifications and tolerances. To design engineers, the bill of materials is simply a list of the parts required to make the final assembly, and the production of the bill of materials involves extensive lettering on drawings and requires time that engineers frequently feel could be better used developing new products or redesigning others. In many organizations the bill is produced by a draftsperson in engineering rather than a designer.

Although design engineering ostensibly supports marketing with new product designs, marketing also needs adequate information in catalogs, manuals, spare parts bulletins, and other documentation. Engineers have no desire and frequently believe they have an inadequate writing education to produce instruction manuals. They also think that illustrating catalogs and manuals is a waste of their technical talents. In short, documentation is viewed as someone else's job. With the emphasis on new designs needed for company future growth, the support of marketing is often less than adequate.

Traditionally, design engineering has been thought to encompass the first four roles listed previously. Problems arise, however, when the last two are added. Although design may agree to support manufacturing, there is usually some conflict over what manufacturing needs for documentation. These manufacturing needs are frequently not understood by design engineers, to whom documentation is the least creative activity in the department. Manufacturing needs designs that are efficient to manufacture, not those that provide the greatest complexity or sophistication. Manufacturing needs products designed for ease of assembly and ones that take advantage of available processes. It needs subassembly definition as well as lists of parts for identification. Some designs may overload work center capacity in some work centers, and a slight modification may eliminate that overload. Design engineers must understand the capacity impacts of their designs. Ivory tower isolation inhibits consideration of these manufacturing needs.

The need for design engineering to support materials control arises from the new role of the bill of materials in modern systems; Chapter 8 will cover bills of material and the role they now play. When order point techniques were used in inventory control, the bill of materials did not figure in material acquisition. Material Requirements Planning (MRP) technology, however, uses the bill to derive component material requirements. This makes the bill of materials the framework for the entire material control function. Design engineers often do not understand these new requirements and, as a result, fail to support material control needs.

The Manufacturing Control System and Material Requirements Planning techniques are capable of improving overall plant productivity, not just that of direct labor. Design engineers can use data from these modern systems in their own activities and must support both manufacturing and material control to achieve the potential systems benefits for the company. Design engineers need to understand the role documentation will play in these systems and to understand how these systems can support their efforts as well. The functions attributed to design engineering to support these roles are shown in Figure 1-1.

Difficulties with other departments surface frequently in executing these functions. Many problems begin with part numbers. Manufacturing needs short, nonsignificant part numbers to reduce reporting errors, as well as part numbers for subassemblies and semi-finished components, which may be of no interest to design engineers. By contrast, design engineers think that significant part numbers containing code descriptions of form, material, func-

Figure 1-1. Functions of Design Engineering

Design new products
  Control: drawing numbers
    model numbers
    part numbers
Issue engineering change orders
Provide drawings
Issue bills of material
Provide material specifications
Provide product specifications
  Compliance to standards data
  Handbooks
  Installation instructions
Provide operating manuals
Specify repair parts
Conduct research and development
Test prototype conformance
Standardize components
Provide technical assistance
Reduce product liability
Specify manufacturing process (process industries)

tion, and group class code information assist their job of standardizing components. Similar conflicts arise in drawing and product model numbering. The real question today is how to get our systems to support both needs. This topic is covered in Chapter 7, "Part Numbering and Coding."

Such conflicts at the start of the manufacturing process are intolerable. These conflicts are avoidable if adequate planning for system support is assigned to the design/manufacturing interface. Creativity is necessary to design; this is a primary reason for isolating design from other functions and is discussed in Chapter 3. Other areas of conflict are discussed in later chapters along with solutions. In today's competitive world companies cannot afford conflicts that reduce productivity and produce bottlenecks or wasted efforts either in design engineering or in manufacturing.

THE MANUFACTURING ENGINEERING DEPARTMENT

Manufacturing engineering is usually assigned the responsibility of determining the producibility of a design, the methods and standards used in production, and the tooling design required to support the process; in short, it is responsible for the way products are produced in the plant. This department might be assumed to be free from conflicts with manufacturing but this assumption is wrong. The roles of manufacturing engineering are to

1. Review designs
2. Provide
   tooling
   routings
   methods
   standards
3. Support manufacturing
4. Support material control

While design engineers are assigned the responsibility of producing the best design, manufacturing engineers are assigned the responsibility of making sure the company can produce it. This is especially crucial for companies with design departments located some distance from the manufacturing processes. By review, manufacturing engineering determines if the design can be feasibly and economically manufactured. This does not necessarily result in the best design but it does yield a producible product.

Manufacturing engineers are assigned the job of providing tooling, jigs, and fixtures to permit manufacturing, inspecting, and testing the product. They are also asked to provide methods and standards by which the manufacturing process is measured for management control. The management and control of direct labor is very important to process control, and thus standards must be constantly reviewed and reevaluated to maintain accuracy. Since these standards become part of the cost data used by accounting and incentive systems to determine standard performance, there is considerable pressure to have repeatable and precise time studies on each operation of every process.

The function of supporting manufacturing includes the choice of new equipment and processes, redesign of existing processes and equipment, and layout of work centers within the plant. These are developed to optimize the manufacturing process for the greatest productivity. Often when there are changes suggested which improve material flow, dissension arises because of supervisory territory, and pay policies. For example, supervisors whose pay is based upon the number of people supervised rarely want their territory reduced. Using cellular groupings of machines with dissimilar functions to produce similarly shaped parts results in faster, smoother flows.

Setup time is dependent upon tool design and equipment used. Manufacturing flow improves by using slower running equipment that is easy and quick to set up. This is counter to the widely held philosophy that large batches are better than smaller ones. This is still taught in most engineering schools and believed by many company managers. Chapter 10, "Lead Time and Setup Time," and Chapter 15, "The Just-In-Time Philosophy," will explore the-smaller-the-batch-the-better concept.

The first three roles of manufacturing engineering are traditional and usually well understood, but the fourth generates considerable difficulty. Manufacturing engineers are usually very familiar with the processes in the factory but rarely aware of how material control and manufacturing are made more productive by their choices for tooling, routings, methods, and standards. Few manufacturing engineers understand the manufacturing control system and the ways in which they can affect it. For example, most manufacturing engineers are familiar with the determination of economic lot-sizes by the square-root formula. One factor, the cost of reordering, includes the machine setup cost. Another factor, the carrying cost, includes the storage cost of the parts. It is not unusual for engineers to understand that setup time reductions will produce lower reordering costs (they can see that in the EOQ formula). Few, however, see clearly the direct relationship between tool design, setup time, lead time, and backlog reduction to reduce reorder cost. In addition, it is often unclear to engineers that the design of expensive quick-change tooling and efforts to reduce setup time will pay easily for more expensive tooling. These efforts will permit backlog (work-in-process inventory) reductions which result in shorter lead times, higher productivity, better customer service, and simplified manufacturing control systems. The savings far exceed the cost of quick-change tooling. The problem is greater than simply getting engineers to understand these principles. They are often prevented from implementing the changes by accounting practices that put a value on low-cost tools and do not measure the benefits of improved flow in the plant.

Tooling budgets are important but the best tooling is not necessarily the most expensive. Frequently ignored and difficult to compute are the benefits of improved flow of materials through operations. Such flow improves with short lead times, running small batches, and using short setup tooling. Often such great emphasis on budget variances, without any acceptable data on the benefits of improved flow, places smooth flow at low priority.

Modern integrated systems for control of manufacturing provide the ability to measure the results of improved flow. Those companies that have educated their manufacturing engineers about the capabilities of modern systems and directed them to attack the problems of high setup times, large backlogs, and big batches have proven the value of better flow. The benefits have far exceeded the costs.

The functions supporting the roles listed above are shown in Figure 1-2.

The manufacturing engineering department, like design engineering, is involved in conflicts which arise between departments. These occur in tooling, numerical control tapes, routings, standards, and methods, and they result in friction with the manufacturing group. Tooling takes too long to produce, delaying the manufacture of needed components. Numerical control tapes start machining in locations other than suitable to match fixtures already available to hold the parts. New routings use already overloaded work centers and stand-

Figure 1-2. Functions of Manufacturing Engineering

1. Review designs for manufacturability
2. Identify tooling requirements
3. Design tooling
4. Provide tooling availability schedules
5. Develop and document
   Routings
   Methods
   Standards
   Alternates
6. Provide numerical control data
7. Develop plant capacity data
8. Maintain manufacturing data base
9. Specify new equipment
   Design
   Data for justification
10. Provide technical assistance

ards and ignore the real complexity and variability of setups. Poor manufacturability of designs produces conflict with design engineering. Differences in valuations of high utilization of plant capacity and tooling costs result in conflict with accounting and management. Expensive machines are not busy enough, the tools are too expensive, and the work centers are frequently unable to produce the work in standard time allowances. These conflicts will continue but can be controlled and used to optimize the whole manufacturing environment only if manufacturing engineering integrates its activities into the manufacturing control system. The system must give engineers the necessary assistance to execute their functions efficiently. This is discussed in depth in Chapter 11, "Manufacturing Engineering."

THE PLANT ENGINEERING DEPARTMENT

Responsibility for the physical plant and maintenance of the plant and equipment are assigned to plant engineering. However, plant engineering and manufacturing engineering share control over plant layouts. Manufacturing engineering has responsibility for equipment location, layout, and material flow, but plant engineering strives for efficiency in power distribution, lighting, utililties location, HVAC requirements, and building construction. Thus, goals may be different and conflict is frequent over layout control. These two departments must work together on process layouts and machine locations, but the layouts themselves should be the property of plant engineering. The overall reasons for this are discussed more fully in Chapter 12, "Plant Engineering."

Plant engineering usually carries responsibility for plant and equipment maintenance, both breakdown and preventive, as well as energy and energy control. Plant engineering usually has responsibility for buildings, grounds, and the maintenance of production equipment, especially since they have the requisite equipment and maintenance personnel. Their entrance into the energy and energy control field is a direct outgrowth of the responsibility for energy distribution and unit control systems in the facilities sphere.

The roles of plant engineering are as follows:

1. Maintain facilities and equipment
2. Design facilities
3. Plan facilities
4. Conserve energy
5. Support manufacturing
6. Support material control

This engineering group provides its greatest support for manufacturing through its maintenance function, at least on the surface. Conflicts are customarily over repair times for equipment or vital facilities, but they also extend to the areas of facilities designs and energy conservation measures. Facilities must provide maximum flexibility of use; power distribution busway changes, for example, should easily permit layout changes to improve flow. Unfortunately, however, facilities plans are often crisis-oriented. Facility planning starts when insufficient space is available for a new process. This results in many small expansions done haphazardly rather than rationally designed, systematically constructed expansions designed to fit the company business plan. This link to the business plan is explored in Chapter 14, "Long-Range Facilities Planning."

The activities which support manufacturing include repair parts inventories and scheduling of tool maintenance. Often machine spares are not available when needed, and tools are unavailable for reuse when they are desperately needed. Manufacturing shares the blame when preventive maintenance is not scheduled; equipment cannot be repaired and run at the same time.

Scheduling maintenance, repairs to tools, and emergency repairs also result in conflicts between the material control group and plant engineering. The scheduling of preventive maintenance causes friction with both manufacturing and production control. Then, too, accounting wants to adhere to the budgeted costs of maintenance and questions how preventive maintenance can be justified. These conflicts are avoidable with the help of properly designed manufacturing control systems which can schedule tool and machine maintenance as well as production. Plant engineering needs as well as manufacturing needs must be considered in the design of such systems. Information is needed by plant engineers on repair parts inventories, tool failure and repair

Figure 1-3. Functions of Plant Engineering

1. Maintain equipment and facilities
2. Provide lubricant specifications
3. Control maintenance inventory
4. Direct energy programs
5. Provide and maintain pollution controls
6. Schedule maintenance staff
7. Determine facilities requirements
    Buildings
    Utilities
    HVAC systems
    Energy
8. Coordinate design of new facililties
9. Manage facilities
    Construction
    Modification
10. Provide governmental reports
11. Provide technical assistance

histories, machine failure frequency, and energy consumption. These data are needed to insure maximum plant efficiency, and the modern manufacturing control system should provide them. These system needs are covered in Chapter 12, "Plant Engineering."

The functions required to support the roles of plant engineering are shown in Figure 1-3.

These functions can be supported by a system that schedules maintenance on production equipment, maintains repair parts inventory records, maintains energy capacity planning data, orders tool repairs based upon usage, and plans for maintenance as a resource need. Plant engineering can make use of the system to keep equipment running to maintain work flow, optimize preventive maintenance, tie plant needs to the business plan, and maintain tooling to produce quality products. These require both a properly designed system and an engineering group educated in system operation. The present system technology is capable of doing these jobs. It needs to be integrated and used by plant engineering to produce results.

In summary, the roles of design, manufacturing, and plant engineering must be integrated rather than fragmented. Manufacturing and material control functions are vital to engineers if they are to be effective in the manufacturing environment. A conceptual look at the manufacturing control system in Chapter 2, "Manufacturing Control Systems," will provide the foundation for the integration of engineering into all other manufacturing activities.

2

THE MANUFACTURING
CONTROL SYSTEM

HISTORY

Only a few of the techniques needed to control manufacturing were available by the start of World War II. The oldest technique is machine loading using a method developed about 1900. Frederick W. Taylor showed how to set standards for industrial work, and Henry Gantt applied bar charts, which combined to give us machine loading data still popular today.

The second tool to become available was the Economic Order Quantity (EOQ) formula published in 1915. This formula balances the cost of ordering material against the cost of carrying inventory to minimize the total. The quantity that produces this minimum is the most economic quantity to order. This technique has seen many enhancements since its introduction: Least total cost, least unit cost, lot-for-lot, part-period balancing, period order quantity, and Wagner-Whiten's algorithm have been added to the standard square root formula.

The third technique, developed in 1934, was Statistical Safety Stocks. This applied probability and statistical mathematics to determine how much inventory to carry to be sure fluctuations in demand didn't result in too many stockouts. Out-of-stock conditions were and still are considered to be one of the deadly sins in manufacturing. A proliferation of fancy formulas to compute the needed safety stock has coincided with the advent of the high-speed digital computer.

These fundamental tools have been expanded, correction-factored, and computed dynamically for use by many manufacturing companies. Other tools, however, are now needed and available. In the late sixties and early seventies a great new tool appeared—Material Requirements Planning (MRP). This technique improved the inventory planners' ability to anticipate material needs and to plan to obtain the needed quantities to produce the product.

This new technique received mixed reactions. Some reacted negatively to MRP and chose not to use it. Some became so enamored with it and its MRP acronym that we soon had MRP, MRP I, MRP II, DRP, BRP, and numerous others. These acronyms have served mainly to confuse our language and understanding. If asked, the creators of these acronyms would probably say that they are really different techniques with a common origin. They all have a similar basis—the technique of Material Requirements Planning, with the emphasis on PLANNING. Planning is needed and better planning yields dividends, but the goal is CONTROL, the control of manufacturing. Better planning alone doesn't produce better control as many seem to believe.

CONTROL SYSTEMS

Control systems are composed of components connected together to provide a predictable result in a process from a given set of inputs. This classical definition fits both the open-loop and closed-loop type systems. The open-loop system is the simplest but the least effective in achieving control in erratically operating processes. Figure 2-1 shows a diagram of an open-loop system which consists of a controller connected to a process producing an output. The desired output response is fed to the controller and the controller issues instructions to the process to achieve the desired response. If the process functions consistently, the output will match the desires given the controller. If the process is inconsistent and erratic, the output will not match that desired. The system is called open-loop because the controller has no way to measure process output. The technique of Material Requirements Planning (MRP) is a mathematical tool to plan order launching, but no feedback exists for output measurement and comparison. MRP, as the name implies, is a material PLANNING tool. This no-feedback defect of open-loop systems is corrected in closed-loop systems, which feed back actual output to the controller so as to

Figure 2-1. Open-loop Control System

Figure 2-2. Closed-loop Control System

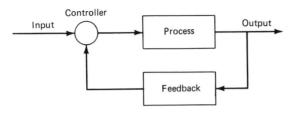

(Courtesy of George Plossl Educational Services, Inc.)

match the desired with the actual. A diagram of the closed-loop system is shown in Figure 2-2.

CUSTOMER DEMAND AND MANUFACTURING

Customers almost never issue orders at the same rate that companies want to make the product. Customers want maximum choice of options on products and frequently seek new options. Manufacturing wants stable plant loads, a constant labor force, smooth batch material flow, and standard products. In short, customers want flexibility and manufacturing wants stability. There are many ways to resolve this conflict. It is possible, in a few cases, to limit customers' choices and make customers live with quantities dictated by schedules at times of peak demand or to provide price and service inducements during slack periods. It is also possible to stabilize customer demand, but the price paid is often high in terms of future business going to the competition. Inventory may be held to buffer demand fluctuations. This alternative is very useful where demand is seasonal, but highly optioned products create high inventories. Another way is to provide flexibility in manufacturing to meet customer demands, as variable as they may be. Costs are associated with flexibility: additional equipment to meet peak demand, transfers, layoffs, and rehiring of labor—along with the increased costs in buildings to house it all. The question is not which to do; the question is how does one strike some balance between what appear as two bad alternatives; it costs something to do either.

It takes managerial judgment to decide what balance to strike between order pattern improvement, flexibility in manufacturing, supporting inventory, and reduced overall profits. Even today there is no way to make these judgments quantitatively or by rigorous computer calculations; good judgment is still the solution. There is hope, however. Complete, well-integrated, formal manufacturing control systems can provide the facts to managers at all levels that will permit them to make more rational decisions. It is essential that managers know what information they can get from these systems and how to use them to evaluate the potential consequences of their decisions before making those decisions.

The manufacturing control system is a feedback control system, but the feedback is information and the controller is company management. The system is not a producer of decisions or control signals but a producer of information fed to decision makers. People provide the control decisions used in the manufacturing plant. Thus, the manufacturing control system is both a PLANNING and a CONTROL system with vital human components—the managers and the users. The system has both a priority and a capacity side working in concert. As with any control system, all modules must be properly connected and functioning for the system to work. Figure 2-3 is a system flow chart of a universal manufacturing control system.

Figure 2-3. Universal Manufacturing Control System Flowchart

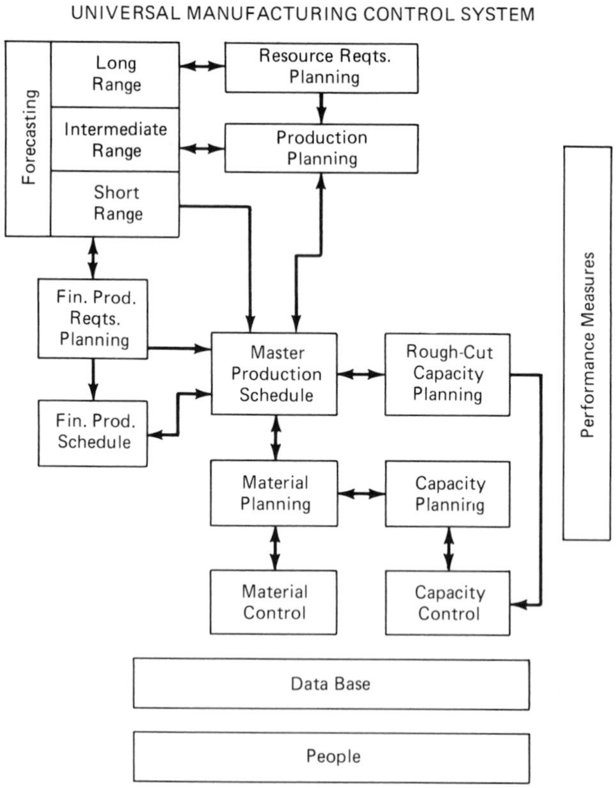

UNIVERSAL MANUFACTURING CONTROL SYSTEM

(Courtesy of George Plossl Educational Services, Inc.)

THE MASTER SCHEDULE

Master scheduling is the process by which management makes a best-guess plan of what they wish to produce over some time period. As such, it includes input from the sales department and from marketing, but the MPS does not show what the company plans to ship—it shows what the company plans to make. The plan horizon and time periods are variable from company to company, depending on their individual needs. The customary period for most companies is one year, but it may be as long as two years or as short as a quarter of a year. Figure 2-4 shows a typical Master Production Schedule.

The Master Production Schedule is the company's course, describing its destinations, course, and arrival schedule for some finite time period. It has nothing to do with a sales projection or maximum production forecast but is instead a realistic plan for production. The Master Production Schedule is the plan that balances the needs of sales, financial management, and manufacturing to produce a quality product efficiently that will satisfy customers' needs. It is the desired result and course to achieve that result in the form of a schedule, analogous to a ship's course chart.

As with most planning, the course specified in the MPS is a *desired* result and is subject to revision. The course chart must be achievable given realistic conditions. A plan based on ideal conditions will not match actual performance, nor will excessively optimistic plans. Thus, the course set by the company in the master schedule is checked carefully to see if it is achievable. Capacity planning techniques do this validation.

The Master Production Schedule shown in Figure 2-4 shows that the company expects to produce 10,000 pens every three weeks over the 14-week horizon. They have decided to produce pencils, 6000 at a time, in weeks 2, 5, 8, 11, and 14; 2000 pointers will be scheduled in weeks 3, 6, 9, and 12, when pens and pencils are not scheduled; and antennas are scheduled to be produced in batches on a schedule of 1000, 2000, skip a week then repeat the pattern. This staggering of production smooths the variation in production loads generated in work centers. Effective master scheduling requires a knowledge of the capacity needed in work centers to support the plan.

CLASSIC AND ROUGH-CUT CAPACITY
REQUIREMENTS PLANNING

Capacity is the rate at which a center, machine, or shop can produce the desired output. A rough-cut capacity plan is a means to assist in schedule development and to check the master schedule's validity. Like all rates, capacity is measured in units per unit time, such as standard hours per week. Backlogs in work centers are measured in units of time, such as standard hours. Great

Figure 2-4. Master Production Schedule

Week No.	1	2	3	4	5	6	7	8	9	10	11	12	13	14	Total
Pens	10			10			10			10			10		50
Pencils		6			6			6			6			6	30
Pointers			2			2			2			2			8
Antennas		1	2		1	2		1	2		1	2		1	13
Totals	10	7	4	10	7	4	10	7	4	10	7	4	10	7	101

Unit of Measure = thousands

Figure 2-5. Bills of Labor

Work Center	Pens	Pencils	Pointers	Antennas
45	10.3	6.5	25.5	25.5
67	7.9	3.1	1.5	1.5
105	8.9	7.0	4.0	4.0
125	1.6	1.2	3.0	3.0
310	3.1	2.0	1.6	1.9

Summary Rough-Cut Capacity Requirements

Work Center	Actual Capacity	Desired Capacity/Week													
		1	2	3	4	5	6	7	8	9	10	11	12	13	14
45	100	103	65	102	103	65	102	103	65	102	103	65	102	103	65
67	40	79	20	6	79	20	6	79	20	6	79	20	6	79	20
105	50	89	46	16	89	46	16	89	46	16	89	46	16	89	46
125	12	16	10	12	16	10	12	16	10	12	16	10	12	16	10
310	15	31	14	7	31	14	7	31	14	7	31	14	7	31	14

care must be used in drawing capacity conclusions from machine loading charts. Chapter 11, "Manufacturing Engineering," discusses loads and capacity in greater detail.

One technique for rough-cut capacity planning is the bill of labor shown in Figure 2-5. It gets its name from its structural similarity to a bill of materials. It is a summary of the work required in total standard hours (or other capacity measure) in each work center to produce a product or product family. Work centers include important vendors; this provides rough-cut planning for vital vendor-supplied items as well. This labor component summary or bill of labor is used to develop a rough-cut capacity plan based upon the master scheduled quantities of the families in each period.

The average outputs, not engineering standards, for each center for each product family produce a bill of labor in standard hours per thousand pieces or some other convenient unit of measure. This bill of labor converts the requirements in the master schedule to desired capacity per week. A comparison of the desired average standard hours per week against the demonstrated standard hours per week will validate the ability of manufacturing to meet the master schedule, on average.

For the four products in the master schedule, the bill of labor in Figure 2-5 shows the average capacity required in the five work centers through which the products and components flow. For example, pen production will require 10.3 standard hours per thousand pens (including all components) produced in work center 45. Pencils will need 6.5 standard hours, and pointers and antennas will need 25.5 each. Multiplying the quantity each week in the master schedule for each product by the average standard hours of capacity needed determines the desired capacity per week. Thus, 10,000 pens in week 1 at 10.3 standard hours per thousand yields 103 standard hours of capacity required.

This technique or a similar technique provides validation of proposed master schedules. It is imperative that master schedules be valid to control manufacturing. If the MPS is unrealistic, the material plans and detailed capacity plans will lead to excess inventory, shortages, high cost, and low customer service—all the things the modern system is supposed to help prevent.

Comparing the actual capacity of 100 standard hours to the average capacity required week by week detects limits. It is easy to see that it is not possible to produce anything except pen components in work center 45 in a week that requires production of 10,000 pens. The capacity of 100 hours is entirely consumed by pen production. By repeating the process for each product and work center with the goal of producing the total volume of product needed within the horizon, leveling the capacity required in each work center every week, an "optimum" master production schedule is produced. This schedule will be valid if and only if the average capacity needed in each work center is at or below the maximum actual capacity available. Scheduled quantities are limited by minimum batch sizes for products so that, in reality, level capacities in work centers rarely occur. It is important to note that the smaller the batch that can be run, the lower the capacity fluctuation will be in work centers and the easier it is to master schedule the plant. Understanding this will be central to the development of the just-in-time approaches to manufacturing discussed in Chapter 15, "Just-In-Time Production."

PRIORITY PLANNING

Priority planning is the process needed to make or acquire the right things at the right time. Priority planning is the process of setting the batch run quantities along with the start and finish dates for manufactured items. To make something, raw materials and supply parts must be available to meet the part need dates for start of manufacture and assembly. There must be an effective ordering system for materials to meet the master schedule requirements if minimum inventories are the goal. Attempting to keep massive stocks of everything to meet these needs would create unmanageable inventories.

It is essential to realize that any priority planning system must incorporate the following in order to be effective:

1. Lead times (averages) must be reliable.
2. Capacity to produce must be available when needed.
3. Materials can be obtained in time.
4. Data on on-hand and on-order quantities must be accurate.

Schedules cannot be met on specific items if the plant is unable to produce what is needed in total. This is true both for plant work centers and vendor plants. This also ties priority and capacity planning together in manufacturing planning systems if control is to be achieved.

MATERIAL REQUIREMENTS PLANNING

Material requirements planning is a priority planning technique that does two things:

1. It nets needs against available inventories.
2. It offsets order release dates based on lead time from the need dates given by the master schedule.

If lead times are not reliable, available inventory and open order data are inaccurate, or if capacity is unavailable to meet the materials plan, the plan will be unrealistic and impractical.

The logic of Material Requirements Planning is as old as manufacturing itself. The essence of the technique is to determine what to build and when (Master Production Schedule) and use the bill of materials for the product to determine the required materials. MRP compares the requirements with existing supplies and computes the net material needs. When our ancestors first made stone axes to cut trees, they knew (the bill of materials) that they needed one axehead, one tree limb, and ten feet of tying material per finished axe. Material requirements planning told primitive toolmakers that if they wanted five axes and had three heads, seven tree limbs, and 40 feet of tying material, they would need two more heads and ten more feet of tying material. This same technique is used by homemakers to make shopping lists to buy groceries for a dinner menu. It is also used by printers to obtain the paper, ink, and binding materials used in this and other books.

For the axe, all materials must be ready for final assembly at the same time. Final assembly takes time so the "when desired" for the five axes is fed into the MRP technique along with the "how long it takes to make" or get the components. The result is a plan to provide the axes when desired. MRP has three functions:

1. Determine the net material requirements.
2. Determine when to start each component to achieve the final assembly schedule.
3. Determine when materials are really needed.

Material Requirements Planning yields these data accurately provided the information fed the technique is accurate. The bill of materials must be correct and complete. The lead times required to obtain components and produce products must be accurate, and the count of materials on hand must be correct. Although the logic is simplicity personified, the operation of MRP in real factories becomes more complicated.

In factories, the inventory consists of raw materials, finished components, and components in process. Second, components and products are made in batches or continuously and not just to satisfy an order. Third, components made are used in several products, and the requirements for all products must be considered. A simple example of material requirements planning in a factory making pens, pencils, telescoping pointers, and antennas is shown in Figures 2-6 and 2-7.

The bills are tree structures with the parent at the top and the components below. Each item has the part number under the name, and the quantity per assembly is one unless otherwise noted. The *500 between Stage 1 (part no. 5134) and the tubing (part no. 6406) denotes a quantity of 500 feet of tubing per thousand Stage 1 pieces.

The material requirements plan for the body (part no. 3440) and the

Figure 2-6. Bills of Material for the Pointer and Antenna

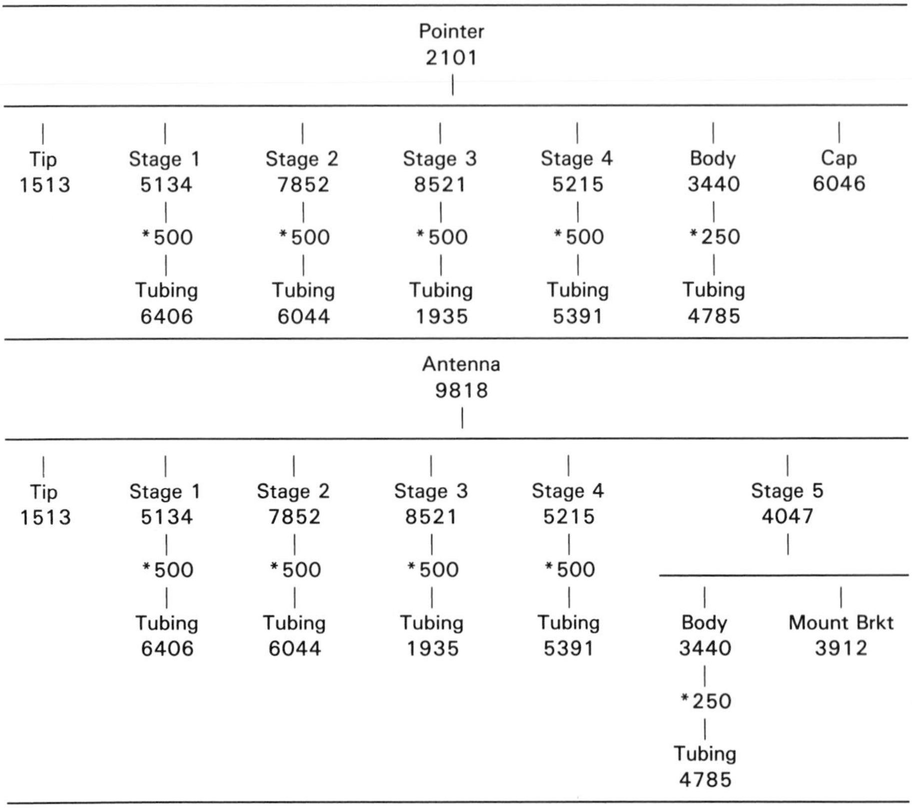

Pointer
2101

Tip	Stage 1	Stage 2	Stage 3	Stage 4	Body	Cap
1513	5134	7852	8521	5215	3440	6046
	*500	*500	*500	*500	*250	
	Tubing	Tubing	Tubing	Tubing	Tubing	
	6406	6044	1935	5391	4785	

Antenna
9818

Tip	Stage 1	Stage 2	Stage 3	Stage 4	Stage 5
1513	5134	7852	8521	5215	4047
	*500	*500	*500	*500	

					Body	Mount Brkt
Tubing	Tubing	Tubing	Tubing	3440	3912
6406	6044	1935	5391		
				*250	
				Tubing	
				4785	

---

* = ft/1000 pieces

tubing (part no. 4785) has a header. It shows the part number, description (Desc), on-hand balance (On Hand), unit of measure (UM), safety stock level in the unit of measure (Safety Stock), order quantity in the unit of measure (Order Qty), lead time, allocated quantity in the unit of measure (Alloc), a purchased (= P) or manufactured (= M) notation (P/M), and an action notice.

The detailed plan below the header shows the week number for week 1 to week 14 preceded by a past due column labeled P/D and followed by a total over the 14-week horizon. The required quantities of the part are shown on the line labeled "Require" in the week they are needed. The scheduled receipts (Sched Rec) show parts due from already released orders. The quantity available appears on the next line. The quantity available before subtracting the requirements for week 1 appears under the column labeled P/D for convenience only. Subtracting the allocated quantity and safety stock from the on-hand balance and adding the scheduled receipts computes the available

Figure 2-7. Material Requirements Plans for The Body and Tubing

| Part No. 3440 | Desc = BODY | On Hand = 48 | U/M = THOUS | Safety Stock = 2 |
| Order Qty = 40 | Lead Time = 8 | Alloc = 0 | P/M = M | Action = NONE |

Week No.	P/D	1	2	3	4	5	6	7	8	9	10	11	12	13	14	Total
Require		5	15	2	0	16	2	5	16	2	0	16	7	0	6	92
Sched Rec									40							40
Available	46	41	26	24	24	8	6	1	25	23	23	7	0	0	-6	
Plan Due															40	40
Plan Rel						40										40

Pegged: 1200 — Pencil Assembly   4047 — Stage 5 Assembly
2101 — Pointer Assembly   4315 — Body Assembly

Figure 2-8. MS/MRP Explosion

Material Requirements Plan

| Part No. 4785 | Desc = TUBING | On Hand = 12 | U/M = THOUS FT | Safety Stock = 0 |
| Order Qty = 30 | Lead Time = 6 | Alloc = 10 | P/M = P | Action = NONE |

Week No.	P/D	1	2	3	4	5	6	7	8	9	10	11	12	13	14	Total
Require							10									10
Sched Rec							30									30
Available	2	2	2	2	2	2	22	22	22	22	22	22	22	22	22	
Plan Due																
Plan Rel																

Pegged: 3440 — Body

quantities. Planned orders are not available until they are actually released and the quantities scheduled as due in a certain week. When the available quantity drops below zero, the system schedules a planned order due (Plan Due) for that time period. It will compute the release date for the planned order (Plan Rel) by subtracting the lead time and schedule the quantity in that week. "Pegged" at the bottom provides the name and part number of the parent which generated the requirements for the part.

Material Requirements Planning logic is very simple as Figure 2-8 reveals. The master schedule is at the top; the plan is to produce 20 units in week 2, 30 units in week 4, 25 units in week 6, and 35 units in week 8. To produce 20 units in week 2, 20 units of final assembly components are required. The on-hand balance is 10 now and 10 more will be received in week 2 so the need for final assembly components has been met. The 30 final assembly components required in week 4 are not yet covered and must be ordered. If it takes one week to get them, an order must be placed in week 3, shown by "start" quantity. This order to make 30 in week 3 generates a requirement for the subassembly's components in week 3. An order has already been placed for 30 and delivery is expected in week 3. This satisfies the need for this component in order to complete 30 units of finished product in week 4.

Now jump to the 35 units planned in week 8 of the master schedule. The requirement for final assembly components has not been satisfied so the need generated in week 8 must be ordered in week 7. This in turn requires that 35 subassembly components be on hand by week 7. With a two-week lead time, the order for subassembly components must be started in week 5—thus the requirement for 35 in the purchased component in week 5. An additional ten units of this purchased component are required each week for spare parts, giving a total requirement in week 5 of 45. The needs for this purchased component for weeks 1 through 4 can be met by the 90 in stock, but the requirement for 45 in week 5 will produce a need for 20 more. With a four-week lead time and an order quantity of 100, an order must be started for 100 in week 1. Thus, the order placed in week 1 for the purchased component is generated by the planned needs for final assemblies eight weeks from now. That is the logic of MRP and the forward visibility it provides.

ORDER POINT TECHNIQUES

Because of the difficulty of keeping up with all these calculations, MRP was not practical without computers. The advent of the computer and the ability to store masses of data and do calculations in nanoseconds made MRP viable as a material planning system. Used alone prior to MRP and still widely employed with it today is the Order Point technique. This technique includes the familiar min/max and two bin variations as well as fixed order quantity-variable order cycle techniques. Plant engineers routinely employ the two bin system for frequently needed repair parts. The main bin supplies parts until empty; then additional stock is ordered in a fixed quantity while the second reserve bin is being used. The two bin system is identical to the order point system when the second bin contains the order point quantity. Engineers are taught order point techniques in school so it is well understood. However, since they do not receive routine training in the MRP technique, they frequently believe Order Point is the technique used to control inventory.

Order points require no bills of material because they deal with each part independently of all others. For items such as hand tools used for maintenance and repair and repair parts such as bearings or belts, whose demand is only weakly related to production hours (independent demand), this is a worthwhile technique. For complex assemblies of parts, where demand for the components comes from the production of parents (dependent demand), MRP techniques are superior.

ORDER POINT FORMULA

The order point quantity is equal to the anticipated demand during the order lead time plus a reserve or safety quantity, which is meant to compensate for difficulties in predicting usage during a period. There are five fundamental elements to determine the level of safety stock:

1. Demand forecast accuracy
2. Forecast lead time length
3. Forecast lead time variability
4. Order quantity size
5. Desired service level

There are statistical methods to determine forecast accuracy, and we suggest using them with some pragmatic judgment to set safety stock levels. A detailed discussion of MRP and order point techniques can be found in *Production and Inventory Control: Principles and Techniques* by George W. Plossl in the bibliography for Chapter 2. This text and others contain a wealth of discussion on material ordering techniques.

Order point techniques have a role in material planning for items with independent demand. MRP techniques are far better for dependent demand. Note that neither order point nor MRP works very well if inventory on-hand balances, open order quantities, and lead times are incorrect. MRP, however, requires accurate bills of material not needed by the order point technique.

DETAILED CAPACITY REQUIREMENTS PLANNING

Once MRP or some other priority planning technique has been used to develop the detailed material requirements by time period, routings, standards, and performance data are used to develop a detailed work center capacity requirements plan. This plan is the rate at which standard hours are required in time periods to meet the master schedule. Once complete, the question, Is average capacity adequate? may be answered specifically. If the answer is No, the master schedule must be revised and the planning procedure repeated until a valid plan is developed. If the answer is Yes, capacity on average is adequate, the plan is good, and execution can start.

INPUT-OUTPUT CONTROL

The control of manufacturing, to a great degree, is control of the rate of delivery and execution of work in work centers in the plant. When work flows in faster than it can be processed, two conditions result. First, the work-in-process inventory increases, resulting in longer and more erratic lead times. Second, backlogs at work centers increase the difficulty in getting valid priorities on orders, and this often results in broken setups and further delays. Input-output control applies to vendors as well as plant work centers.

Control means governing the rate of order release to the shop and monitoring the rate of work completion to achieve a balanced flow. This balanced flow controls the level of backlogs; the more closely the backlog level approaches zero, the more the batch manufacturing operation appears like a process plant.

Input-output control can be effective only if capacity is adequate. Once maximum capacity is reached, the addition of work at any greater rate only increases lead time and backlogs. Excessive capacity inevitably results in excessive inventory because manufacturing management does not tolerate idle people or machines. Keeping capacity busy results in low cost, excess inventory, not a real benefit. Inventory not needed represents wasted material and capacity that could have been used to produce something that was needed. It is vitally important to balance capacity to maintain control. Thus, input-output control provides effective control of capacity and the utilization of that capacity. It also controls lead times. This technique provides an ongoing check on capacity being adequate all during the execution of the plan so it is truly a control technique.

PRIORITY CONTROL

Once a valid plan has been made, the priority or dispatch list will indicate the sequence to be followed in processing orders within each work center. It is really the production schedule that says, "Make this one first and that one second.'' This sequenced job-by-job dispatch list indicates the run order for specific jobs and determines the capacity required in work centers to complete them. This now makes it possible to test if the specific capacity of each center is adequate to keep orders on schedule. If capacity is not adequate, the material plan will have to change and if the change is drastic it may require a revision to the master schedule. When the answer to the question Is specific capacity adequate? is Yes, the plan is achievable and controllable.

PRIORITY AND CAPACITY

The manufacturing control system is split into functional sections—priority and capacity. The priority section contains the planning techniques—master scheduling and material requirements planning, plus a control module, the dispatch or priority list. The capacity section contains two planning techniques: rough-cut capacity planning and detailed capacity requirements planning, plus a control module, input-output control. The manufacturing control system could be diagrammed as shown in Figure 2-9.

The manufacturing control system consists of four functional groups: priority planning, capacity planning, priority control, and capacity control.

The ability to control capacity and the complexity of the system required is a function of decisions made in engineering. For example, let's look at the effect on the economic order quantity of changes in setup time. The economic order quantity is commonly computed by using the square root formula:

$$EOQ = \sqrt{2AS/I}$$

where   A = annual usage in $
        S = ordering cost (mostly setup cost) in $
        I = inventory carrying cost as a decimal fraction per $ of average inventory

Figure 2-9. MC System Diagram

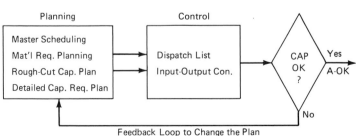

The equation and its absolute evaluation are not important since we wish to examine the change in EOQ as we change S. Rearranging we get:

$$EOQ = \sqrt{2\,A/I} \times \sqrt{S}$$

This means that the economic order quantity decreases as the square root of the setup cost. Since setup cost is proportional to setup time, the setup time must drop by a factor of four to halve the EOQ. This says that dramatic reductions in setup times are required to have any real impact on EOQs. Dramatic reductions take rethinking of the whole setup situation. Working harder, for example, may result in 10 percent reduction in setup. It takes considerable ingenuity and engineering to reduce setups significantly.

Unless engineering realizes the effect that EOQs have on work-in-process inventory and lead times, lack of attention to reducing setups will result in surging plant loads. Instead, engineering effort must be directed toward making work flow more smoothly.

MANUFACTURING CONTROL

Control of manufacturing requires a closed loop feedback control system that supplies people with information needed to manage the processes of manufacturing. Like any control system, it can only provide output within the limits imposed by the physical processes themselves, and no amount of tinkering or sophistication in the software can possibly change the realities of the real plant environment. Engineering controls the design and maintenance of the product, the process, and the facilities. Manufacturing can be planned and scheduled by systems but controlled only by direct application of engineering to make the processes produce efficiently a high quality, reliable product designed for manufacturing. Engineering really controls manufacturing.

Manufacturing control is the body of knowledge related to all activities in planning and controlling manufacturing operations. These activities include everyone in the company from the president to the janitor, and they include computerized as well as manual systems. Manufacturing control includes *execution* as well as *planning*. Execution of the plan produces the product and provides the income from which profits are derived. Improving planning is needed, but engineering can make a great contribution by improving the company's ability to **execute** the plan.

3

DESIGN ENGINEERING

The word *creativity* is one of the more difficult terms in our language to define. Although creativity can be easily recognized, few can provide an acceptable definition for it. The dictionary defines creativity as "the state or quality needed to cause to come into being, as something unique that would not naturally evolve." The term *invention* is often used in conjunction with creativity, but it too is difficult to define. The United States courts and the Patent Office have been trying to develop a workable definition for years. So, although we all understand what creativity and invention are, we have a difficult time developing a workable definition. This book uses the broad-sense definition that creativity embodies the innumerable inventions and design decisions which, when combined, produce a new device, concept, or other tangible work.

Technical creativity falls into one of three classifications:

1. Scientific theory: This kind of creativity is the domain of scientists and some engineering scientists. New theories are difficult or impossible to make happen on schedule, but rather evolve from work done over long periods.
2. The scientific frontier: Here new knowledge is analyzed, explored, expanded, and made usable in society by engineers and scientists, who work together to apply theory to solve practical problems. This effort, like that of scientific theory, is difficult to schedule because new knowledge frequently leads to more questions.
3. New devices, systems, and structures: This is the realm of design engi-

neering, where scientific theory, mathematics, economics, and social knowledge merge to develop products that satisfy specific needs. Although highly creative, this activity can and indeed must be scheduled for successful manufacturing.

CAN YOU SCHEDULE CREATIVITY?

Thomas Edison once said, "Genius is one percent inspiration and ninety-nine percent perspiration." The one percent cannot be scheduled but the ninety-nine can be. Witness to this are our immense efforts in two well-recognized areas: the development of the atomic bomb through the Manhattan project and the Gemini mission to put the first man on the moon. These projects were scheduled and completed on time. Creative efforts can be successfully completed without unlimited funds as well.

It is necessary to manage resources to accomplish the task, not to just throw resources at problems. Excellent project management makes successful projects, not simply ample funding.

To manage effectively, it is necessary to understand the people to be managed. Psychologists researched the characteristics of creative people. The following traits are accepted as common:

1. Self-sufficiency and self-direction: Creativity requires self-starters. They must work on problems because they want to make discoveries. This internal motivation drives the creative personality.
2. Preference for mental challenges: The challenge of grappling with unsolved problems is a hallmark of creativity. It is routine for creative people to have mentally challenging hobbies and avocations.
3. High Ego: Creativity is self-centered. The creative individual is frequently a zealot who disagrees with the current thinking. This individuality occurs in those with a strong ego.
4. Preference for exactness: Creativity flourishes in an ordered environment. The orderly process of sifting, combining, rearranging, transposing, and modifying ideas occurs most frequently in an organized atmosphere. Precision in thinking and simplified, exact results are common to these people.
5. An appreciation for abstraction: Creativity manifests itself in the development of general principles or concepts from specific cases. The ability to generalize from the specific and the never-ending pursuit of general principles is a characteristic of creativity.
6. Preference for isolation: Isolation is a primary defense mechanism for creative people. Isolation provides an environment where creative people are most comfortable.

7. Independent thinking: Creativity requires a rejection of conventional ideas and a perpetual search for alternatives. This independence often accompanies a general disregard for group pressures to conform.

8. High personal dominance: Creativity flourishes in a controlled personal world. Control of the surroundings and elimination of controversy are characteristic.

9. High self-control: The greatest achievements in creativity occur when individuals direct mental activity. The ability to direct activity and to control the mind and body is prominent.

10. Superior intelligence: The ability to rise above conventional thought and synthesize new concepts is a sign of a high IQ. The ability to retain information and acquire new information quickly is also characteristic. However, high intelligence is not synonymous with high creativity.

These traits provide the manager with an understanding of what motivates creative individuals. They also provide cues for managers in fostering the working conditions most appropriate for this type of individual. The key to managing the creative individual is providing direction and scheduling control while at the same time fostering independence.

THE DESIGN PROCESS

The design process itself consists of six distinct phases:

1. Problem definition
2. Invention
3. Analysis
4. Decision
5. Iteration
6. Documentation

These phases find their parallels in the creative process, which is outlined below. We discuss each phase in turn to determine how they can be controlled, guided, and expedited.

a. Preparation
b. Concentration
c. Incubation
d. Inspiration
e. Verification

The *problem definition* is a statement telling the designer what to design. It may be as broad in scope as "put a man on the moon," or it may be as narrow as "design an electronic heart pacemaker." This first step in the design process is critical to project scheduling. The problem must be carefully defined to determine its scope. This definition process not only provides scope but also some measure of complexity.

The problem definition starts the first part of the creative process, called preparation. This section of the creative process is basic research. Although estimates are difficult, this activity can be allotted a rough block of time. Since complete, exhaustive research today is impossible with the volume of information in the world, the designer must concentrate on the major sources that can be tapped in a short time. If the schedule is set up to allow a reasonable research period, the designer can obtain a working knowledge of the field within those confines. This doesn't constrain the creative person unreasonably but it does order the environment.

The next phase of the design process is *invention*. Invention starts with the second part of the creative process called concentration and encompasses the third and fourth stages, incubation and inspiration. Most people believe invention cannot be scheduled, yet scheduling does not mean that a phase will start on Tuesday at 10:30 a.m. and proceed to conclusion in one week and two days on Thursday at 2:30 p.m. Rather, a schedule is a plan or best guess as to the time required. We would like projects to stay on schedule but they rarely execute according to plan. Errors, roadblocks, illness, crises, and similar unexpected difficulties make plans inaccurate. Forecasts are always wrong but that doesn't stop making them useful. Forecast a reasonable block of time for this phase and make it part of the overall schedule.

The third phase of the design process is *analysis*. This phase is not strictly a phase but the start of a series of operations which analyze, revise, re-analyze, and revise again. This looping process optimizes designs. Reviews determine design practicality, reliability, cost, and manufacturability. Analysis also precedes a decision to continue the project. Viable designs at review time determine if the project will continue.

The *decision* phase in the center of the analysis loop provides an exit. Analysis and redesign cannot go on forever if a project is to proceed to manufacturing. Perfect designs are unknown and the overall project redesign time is limited. Designs must be frozen and released to manufacturing or scrapped.

The *iteration* phase starts with the decision to continue. This is sometimes the longest phase because many unforeseen problems arise. Although each solution requires creativity, there are still deadlines to be met. It is in this phase of the project that Computer Aided Design (CAD) offers the greatest help. CAD improves the ability of engineering to analyze and redesign quickly, making analysis more thorough and the final design better and more reliable. CAD/CAM is indispensable in controlling design/development projects.

The last phase of the design process is final drafting, which is also the start of *documentation.* Design documentation enables others to use the assembled data to produce the manufactured product. This phase of the project interfaces with analysis if a CAD system is in use, which in turn yields even higher productivities and shorter design cycle times. Unfortunately, the documentation phase is last and is sometimes considered an uncreative, menial task for engineering.

Poor performance on documentation, especially on bills of material, is intolerable today. Modern manufacturing control systems use the bill as the framework for the planning system. A manufacturing organization cannot permit incomplete or haphazardly prepared documentation if production is to occur in a smooth and controlled fashion.

The documentation needed by manufacturing includes part drawings and bills of material. The modern manufacturing control system requires accurate bills for material planning. The structure or organization of the components on these bills is so important that an entire chapter (Chapter 8, "Bills of Material and the Manufacturing Link") is devoted to this topic. One way design engineering can assist manufacturing is to provide preliminary release of data on those components with a long lead time. For example, engineering can supply some preliminary information about the cast housing for a gear box before finishing the final drawings. The approximate size, shape, number of machined surfaces, and the volume and material specifications are known before the final prints are drafted. Thus material control and purchasing can make preliminary data available to a casting supplier so that raw casting compounds and metals can be acquired before getting final dimensions. Material control can also explore the impact of the new casting on present machining capacity. The point is that preliminary data make estimates of gross requirements possible.

MANAGING THE DESIGN FUNCTION

To manage the design function to achieve optimal results, one must keep in mind the nature of the creative engineer and also treat design/development efforts as projects. Numerous weapons and military systems projects over many years prove that scheduling design/development projects works. The "creativity cannot be scheduled syndrome" is fallacious. The fundamentals of project management are:

1. Determine the tasks to be performed
2. Assign the tasks
3. Get schedule and budget estimates
4. Set overall schedule and budget
5. Identify project review points
6. Follow up

ENGINEERING ORGANIZATION

In 1957, C. Northcote Parkinson wrote his famous book, *Parkinson's Law*. Although the book is hilarious, Parkinson makes some serious points about organizations. His law, simply stated, is "work expands to fill the time available for its completion." Even though most managers are aware that engineers not pursuing essential tasks will pursue nonessential ones, it seems that a large number of engineering organizations follow the law. Preventing wasted time requires an organizational structure. With it, management can monitor and control the activities of its engineers as it does with other workers.

There are essentially four basic types of structures among engineering organizations: project, functional, hybrid, and matrix.

The decentralized corporate organizational structure and the *project structure* are much the same. In the corporation, the chief executive is at the top with plant managers in charge of each plant. Each plant has similar departments with functional duplication between plants. The engineering department project structure is also fairly shallow. In that structure three levels exist: the chief engineer at the top, the senior project engineers under him, and the rest of the organization below them.

The major benefit of this structure is that project responsibility is clear. The major disadvantage is that expertise may be duplicated under each senior engineer as it is in each manufacturing plant. Figure 3-1 diagrams this structure.

The *functional structure* is perhaps the most common and follows directly from the principle that specialization produces the greatest expertise. Although the advantage of this approach is evident, a note of caution must be sounded. The well-known humorous comparison between a generalist and a specialist describes the problem well. The specialist learns more and more about less and less until he knows everything about nothing, whereas the generalist learns less and less about more and more until he knows nothing about everything. In manufacturing organizations, specialists are an asset only if they produce manufacturable designs. Excessive specialization is fatal to manufacturing if specialists operate autonomously. The greatest weakness in the functional structure is the tendency to isolate rather than integrate specialists. Figure 3-2 shows the functional structure.

The functional structure makes full use of specialists without the redundancy of the project structure. However, the specialist can easily become so absorbed in special techniques that the design project goals as a whole may lose their priority. The functional structure also suffers from the disadvantage that it is difficult to pinpoint responsibility, and thus important details may be overlooked. A frequent omission is that a product must be efficient to make as well as functional.

The *hybrid structure* takes advantage of both the project and functional structures and eliminates the disadvantages. This structure has functional

Figure 3-1. The Project Organizational Structure

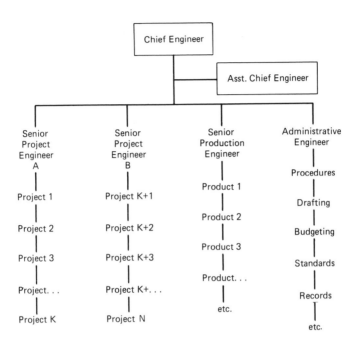

groups to retain and encourage specialization and also has senior project engineers or project managers to handle the flow of projects. Although potentially better than the two previous structures, there is a real danger that this organization will fail in practice. Failure comes when managers have insufficient control over specialists assigned to their projects. Specialists with many projects have time divided among projects and several bosses. This generates conflicts over project priority and in turn produces disharmony. This weakness is potentially more serious than the disadvantages of the first two structures. Unless managers in this environment communicate well and have valid resource plans, organizational failure is the inevitable result. The hybrid structure is shown in Figure 3-3.

The fourth structure, the *matrix,* is a fairly new concept which includes more than just the engineering department. It consists of any number of integrated teams that act as task forces. When individuals cooperate as a task force on a project, they formulate the entire project plan as an integrated whole. Since all functional groups are represented on the team, the matrix approach theoretically produces better, more manufacturable designs than the other structures. Also, because it simulates the small company environment, the project manager must be in firm control of the team to resolve disagree-

Figure 3-2. The Functional Organizational Structure

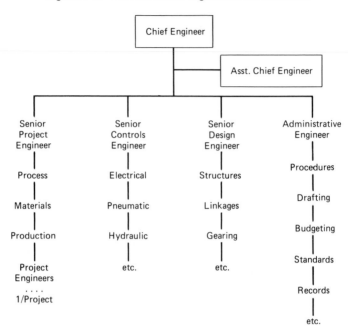

ments. This is the most efficient structure for maximum consideration of all points of view, but to be successful it takes a very good project manager to schedule and direct the conflicting viewpoints. Figure 3-4 illustrates this type of structure.

The task force contains representatives from

Design engineering
Manufacturing engineering
   Methods and standards
   Routings and work flow
   Tooling
   Sales and marketing
   Material control
   Purchasing
   Manufacturing
   Warehousing and shipping
   Plant engineering and maintenance
   Accounting

Figure 3-3. The Hybrid Organizational Structure

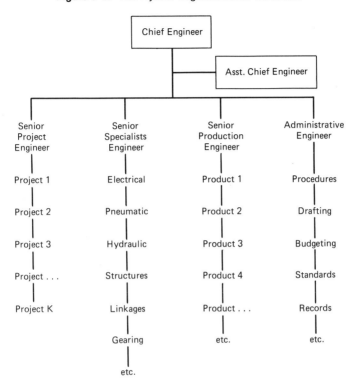

Task forces vary during the project life cycle. Meetings are scheduled for the entire team on a periodic basis but subteams are employed on specific project tasks. Overall project control comes from the project manager, but individuals report to their respective departments as well. Task control and priority are under the control of the project manager because department supervisors assign individuals to the team.

DESIGN FOR MANUFACTURABILITY

The real design goal is to develop a product that is the most efficient to make. This has been sidetracked in many companies as they have evolved into large organizations with increased departmental specialization. Every plant needs

Figure 3-4. The Matrix Organizational Structure

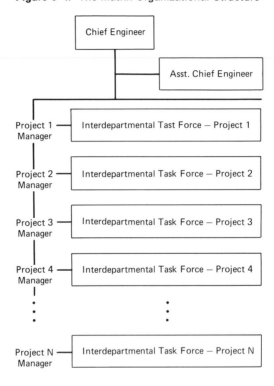

design engineers who understand the manufacturing process. Engineers need to spend time on the manufacturing production floor working on manufacturing problems. This is the exact opposite of what many large companies do. In many companies, design engineering sits in another building or another city far away from production.

Many firms have separate engineers called liaisons. They communicate between engineering and the manufacturing plant. This separation of functions and degree of specialization creates a design elite that understands none of the problems of manufacturing. Manufacturing and engineering must get back to basics and design a product that optimizes the use of the equipment resources available in the factory and can be sold at a profit. That is the real goal of design engineering. Many engineers design sophisticated products that do not use existing equipment and facilities. It takes a knowledgeable, creative engineer to design a product that makes optimal use of the manufacturing process and simultaneously meets market objectives.

Since the design engineer works closely with marketing in new product development, this tends further to separate design from manufacturing. In most manufacturing companies, the design function spends a high percentage of its time developing products similar to existing designs. This is helpful to both engineering and manufacturing in reducing the design time and in producing accurate schedules. For example, in a company that makes heavy machinery, the housings and access doors all require hinges and latches. Engineering can specify a door assembly that contains the door and all the required hardware as a module. This module concept can also be applied to doors in buildings, windows, electrical outlets in homes, and many other similar assemblies. Modules save design time, improve standardization, and provide more manufacturable designs.

Design engineering can be a great help to marketing as well as manufacturing if design engineers spend time in the manufacturing plant. They need to explore the variety of possible products in terms of the processes available. This is opposite to the normally applied method of design developing the product and manufacturing engineering attempting to determine a way to make it. In addition, when the design and processing functions are compartmentalized, each resists change once the details have passed on to the next group. If the design engineer looks at the processes in the plant to determine the variety of possible products, marketing gets product ideas that fit manufacturing capabilities. Design benefits by developing more manufacturable products. In those companies which have tried this approach, the process changes design engineers have suggested have resulted in improved production and in modifications to existing designs which reduce cost, improve productivity, and smooth plant material and production flow.

Engineers in the twenty-first century have a multifaceted tool at their disposal—the desktop computer, which does routine calculations, writes reports, develops budgets, does project analysis using PERT, CPM, or PRRP, and provides a data window into large data bases. As systems get smaller and more inexpensive, desktop computers will become as common a tool as the slide rule of the 1960s. These small computers may inadvertently create one problem that slide rules could not. Although computers can reduce complex tasks using massive amounts of data from many sources into a simple report, the data output, while highly precise, may also be inaccurate. This is because a stand-alone microcomputer can operate on fallacious data very easily, and output is totally dependent on accurate input. The integrated microcomputer and mainframe system that permits data from the business mainframe to be shared with the micro is more likely to get consistent input data. Accuracy is still at the mercy of the data in the mainframe, but we assume that such data are up to date and only one data base needs maintenance.

Another tool available is CAD or Computer Aided Design. This is an extremely powerful tool especially when combined with finite element modeling and analysis software. The problem with CAD systems is the same as for microcomputers—they must interface with the business mainframe or two synchronized data bases must be maintained. It is practically impossible to keep two data bases synchronized with manual data entry; and even tape file transfer, although better, isn't timely in rapidly changing environments. The engineering computer bill of materials must match the bill of materials in the business computer in order for purchasing and manufacturing to obtain and make the correct parts. The CAD system is extremely useful for the design refinement iteration phase, and it is even more powerful when it optimizes the design considering manufacturing process flow. For this to occur, the routings must be available to the CAD system.

Computer simulation is another capability of today's systems. These tools will either revolutionize or set engineering back decades in the manufacturing environment. The revolution will come if engineering makes use of integrated computer data from the business system. The setback will come if engineering with its own computer system isolates itself from the other computerized functions. The football halfback has to learn the play from the playbook to function with the team; engineering is no different. Engineering needs to learn the manufacturing play from business management to be an effective part of the team. The playbook for business is in the business mainframe.

There are many ways that engineering benefits from integrated systems. In design engineering, the business computer contains parts data and bills of material for all items produced. Using these data to help standardize parts facilitates the design process. More on this topic is contained in Chapter 7, "Part Numbering and Coding," and in Chapter 11, "Manufacturing Engineering." This data base can also store bills of material normally put directly on drawings. With everyone using the same data, it is easier to find and correct errors. The results are improved communications, smaller drawings, increased accuracy for the bill of materials, and shorter design time to produce finished drawings.

In manufacturing engineering, the routings used in manufacturing provide valuable data for analyzing material flow patterns. These patterns are shortened and straightened to provide a smoother flow of work in the plant. Capacity requirements data in the business computer also show bottleneck work centers which manufacturing engineering needs to help remove. The rough-cut capacity data can also be used to do sensitivity analyses of material flows to equipment layouts for various product mix changes. These are invaluable for optimizing plant work flow and equipment layouts. If the computer data for routings to run the business and for engineering are identical, the data will have the greatest chance of being accurate.

Plant engineering needs long-range planning data from the business computer for facilities planning. This assists in formulating building requirements and schedules that fit with the business plan. When maintenance is tied to production time and the business computer, equipment can be scheduled for preventive maintenance based upon usage. The computer also tracks mean time between failures for scheduling rebuilds or overhauls. Standard use parts, such as belts or bearings, can be forecast based upon production plans so they can be available for preventive maintenance.

All the engineering departments need data from the business system to improve their efficiency. The business needs engineering data for energy planning and control, bills of material, specifications, spare parts inventory levels, routings, tooling, production standards, and quality management. There seems to be no effective way for engineering to operate with its own computer system independent from the business computer system. Since both groups are on the same team, they both need the same playbook. There seems to be great similarity between successful football teams and successful companies; they function as a team. Communications is key. When management and engineering each begin to discuss their respective problems and needs, the solutions become more obvious and easier to implement. Buried under mountains of inventory and massive, unmanageable data systems, communications can cease and the business can be undercut by the leaner, more cost effective competitor. The first rule of computer systems is that they can become increasingly complex until the added sophistication makes the programs unmanageable. The best computer program is really the simplest. Simplicity is achieved by integrating functions, not fractioning them. Integration is necessary if companies are to remain competitive in the marketplace, and it starts where the product starts—in design engineering.

A good example of how computer potential has been utilized is provided by an electronics company in Fort Worth, Texas. The board production facility has been completely automated. Design and manufacturing work closely together. This has resulted in standard board designs, standardized integrated circuit placements, standard components, and a flow processing line. A bar code on the board is read automatically to determine which program to use to load board components. This bar code is also used by the automated testing area to determine the proper test sequence, the input signals required, and output values to measure. Boards vary in size but use a common carrier with multiple numbers of boards per carrier. This is an example of manufacturing by design. Such a level of automation—permitting production runs as small as one carrier (one to six boards)—is only possible with close cooperation between design and manufacturing.

4

COMPUTER SYSTEMS

Today the computer is as much a part of our engineering environment as a slide rule or calculator was 20 years ago. With the advent of the microprocessor on a single integrated circuit chip, the term microcomputer was born. Engineering schools today find students using powerful hand-held programmable calculators. Many engineering students have personal computers as powerful as some ten-year-old mainframes. The day of the slide rule has passed. Students today have hand-held calculators with continuous data memory, program memory, multiple data storage registers, automatic memory allocation, several levels of subroutines, program flags, multiple subroutine labels, root extraction, and automatic integration. This is on top of over 35 mathematical and conversion capabilities available with, at most, two keystrokes. Twenty years ago this capability was available only using FORTRAN programs on a time-sharing computer that filled a room. A computer with several times that power can now fit in a briefcase.

THE EVOLUTION OF COMPUTERS

Electronic computers have been around for over 30 years. The first real computing machine was the UNIVAC 1 built in 1950. This machine used vacuum tubes and filled a room yet had less capacity than personal computers today. In 1965, Digital Equipment Corporation introduced the PDP-8 computer. For the first time, computing power became available to the laboratory and manufacturing plant production environment. This computer was the first minicomputer. Virtually all computers before this time were called mainframes because they were large, heavy machines that required a substantial frame structure to support all the parts. The minicomputer was small and lightweight compared to the mainframe.

The microcomputer world was born with the development by Intel of the 8008 logic device in 1969. This device was the first single silicon chip microprocessor. This was an event that has had a profound effect on the computer industry. Single chip, complex circuit technology made the personal computer possible. From these early beginnings, Intel developed the 8080, 6809, 6502, 8088, and 8086 processor chips. Zilog developed its Z-80, and Motorola introduced the 68000 shortly afterward. These chips are some the processors used in modern personal and small business microcomputers.

These early microcomputer developments were built around the 8-bit or 1-byte data unit. It was not long after these developments that the 16-bit or 2-byte processor was put on a single chip. Almost overnight the definition of minicomputer was confused by machines using the 68000 processor and the old Digital Equipment Corporation PDP-8 and PDP-11 minicomputers. The definition of a microcomputer as a computer having an 8-bit memory word was now obsolete. The 68000, single-chip processor had 16-bit input/output and did 32-bit arithmetic. Today the best definitions available for mainframe, mini, and microcomputers relate to physical machine size. A microcomputer can fit in the trunk of a car, a minicomputer will fit in a broom closet, and a mainframe requires a room.

DIGITAL COMPUTER CHARACTERISTICS

It is worthwhile to discuss the characteristics of these computers because applications of computer technology for engineering and manufacturing influence equipment selection. All computers regardless of memory word size perform operations one at a time. In the 8-bit microcomputer, each instruction occupies eight bits or one byte, and the processor speed is around four million cycles per second. This means that the 8-bit processor can execute four million instructions per second. The 16-bit processor running at the same speed does two instructions per operation. Thus, the 16-bit machine is twice as fast as the 8-bit. The 32-bit and 64-bit processors do four and eight instructions per operation, respectively. These machines are four and eight times the speed of the 8-bit machine.

Speed, however, is a misleading measure of computer performance. The software or instructions provided to the processor for execution have more influence over execution time than machine memory word size. Some 8-bit machines run fully compiled benchmark programs faster than some 16-bit machines running the same programs. Many 8- and 16-bit computers run the same programs at essentially the same rate. Thus, word size alone means little in terms of program execution time. It is the combination of word size, clock rate, and software that produces fast execution times. The clock rate is the clock signal pulse rate. The pulse signal is usually produced by a quartz crystal oscillator and fed to the processor. In general, the greater the word size capacity of the processor, the faster a given program will execute. It is not safe to assume that a 32-bit processor will run the same program four times the speed of the 8-bit, however. Chapter 5, "Computer Programming," covers programs and program languages more fully.

COMPUTER USAGE CATEGORIES

Computers fall into general usage categories based upon word size. Fixed-point process control systems use the 4-bit or 1/2-byte (called a nibble) computer chips. Boiler controls, some air handling systems, and manufacturing process controls are typical examples. Personal computers use 8-bit processors. The 8-bit machines are programmed to do calculator and word processing functions with some fairly sophisticated mathematical techniques added. Small-size data file manipulation is efficient on 8-bit computers as well. These machines are an engineer's slide rule, calculator, report generator, planner, and tool for mathematical analysis. These machines handle small data files well, but they are inefficient for the massive data bases required for graphics and complex analysis. Processing is also too slow on 8-bit computers for medium-sized business number crunching. The 16-bit processors handle the complex functions such as finite element analysis, computer graphics, and larger file manipulations. The 16-bit processor is available in the personal computer for those with larger tasks. The 32- and 64-bit and larger systems do massive file manipulations and are suitable for large business applications. For the engineer with a frequent need for finite element analysis, a full 32-bit desktop machine is available today.

The main purpose of mass data storage is to keep data available for access by the processor. Tape drives store data in sequence on magnetic tape, which means that searches for data are consequently slow. If access to the data is infrequent as, for instance, archived records, then this storage medium is both adequate and cost effective. The most common type of rapid access, mass storage device is the disk drive. In most personal computer systems this disk drive uses a flexible magnetic medium called a floppy disk. Some 8- and 16-bit systems and all large business computers use the hard (nonflexible) disk. These hard disk drives provide fast access to large volumes of data. The average access time on modern drives is under 36 milliseconds. This is slow compared to direct processor memory access times under .7 microseconds but fast relative to tape access times in minutes. A normal D size drawing on a CAD system occupies between 250,000 and 500,000 bytes of storage. Modern 80-

megabyte hard disk drives can store and retrieve quickly about 320 such drawings.

COMPUTER MEMORY

Many people consider the main processor memory part of the processor itself because of its proximity to the processor. In the early 1960s, memory was usually of magnetic core type. Thus, the term *core memory* refers to main processor memory and not simply the type of memory employed. In fact, there are four types of main memory available today. Magnetic core memory is useful where power failures are frequent and it is crucial to retain memory variables. Numerical control machining operations are a good example. NC machines need to resume operations from the point at which they stopped when the power was lost. This memory permanence costs the user in speed and dollars since core access times are around 1 microsecond and cost $100 per kilobyte.

The second type of memory is metal oxide semiconductor (MOS). Its access time and cost are low at .7 microseconds and under $65 per kilobyte. This is the most common type of memory found in personal and business computers. It is fast and inexpensive but when the power is lost the data disappear as well.

The third type is bipolar memory, commonly called "cache," which has extremely fast access times of about .3 microseconds. Bipolar memory costs around $600 per kilobyte and thus is used where very high speed is necessary. Cache memory is useful for improving throughput time in complex analysis as, for example, in finite element analysis and for some computer-aided design processes. Like the MOS memory, data are lost if the power fails.

The fourth, bubble memory, is now available. Like magnetic core memory, it retains data when power fails but is slow. Bubble memory is cost effective only in rather large capacities. Bubble memory is sometimes a substitute for a disk drive. It has the advantage that it has no moving parts and it takes very little space. Speed is also comparable to that of a disk drive.

COMPUTER PERIPHERAL DEVICES

The peripheral hardware types associated with computers are the I/O or Input/Output devices. These devices get the data and programs into and out of the processor. Perhaps the first device developed for getting masses of data in and out of the computer was the card reader and punch. The reader and punch processed cards at approximately 700 per minute. Although this is slow by modern standards, the reader and punch did a good job of entering and delivering data. The warning "Do Not Fold, Bend, Staple or Mutilate" describes well the difficulties with cards. Cards had to be nearly new for them to work properly. Any small flaw would produce errors. Although still in use today, cards are disappearing and other faster and easier methods are being used.

The second and third oldest I/O devices are the teletype terminal and the line printer. The old teletype machine was both an input and an output device because data could flow both ways at about 100 words per minute. The teletype machine remains one of the more reliable devices available for input/output. Few typists were faster on input but output was too slow for most business applications.

The line printer delivers data from the computer to paper one line at a time—thus its name. Six-hundred-line-per-minute printers are commonly employed, but today's line printers can run at speeds exceeding 8000 lines per minute. Computers normally receive small batches of input data and deliver voluminous output such as summaries, management reports, and charts. The printer was the first device designers improved because computers were slowed by the rate at which data would print.

The most common input/output device today is the Cathode Ray Tube (CRT) terminal. The keyboard permits data input at typing speed, and the CRT provides visual input verification and output displays. Keyboards have standard typewriter keys and many also have ten key pads for numbers and a number of special function keys as well. CRTs are of several different types. The first type is the refresh display or directed-beam display used primarily for CAD graphics applications. This tube draws pictures on the phosphor coating of the tube face with a directed beam of electrons. Since image illumination decays with time, the picture refreshes about 30 times per second to eliminate flicker. These CRTs require considerable memory to store the image and special display processors to refresh the image. Their advantage is that they can instantly change or modify images and even accommodate moving images. This type of tube, like its cousin the storage tube, shows clear, uniform lines and true round objects. Circles drawn on a refresh tube do not look as though they were made of small horizontal and vertical rectangles. Mapping and graphics applications use this tube to eliminate jagged-looking diagonals and curved lines.

The storage tube is similar to the refresh in that the image is painted by an electron gun. The difference is the image resides on a wire mesh and the screen is illuminated by a "flood gun." This type of tube does not need refreshing and thus eliminates the need for memory or processor time to refresh the image. The main advantages are lower memory requirements than the refresh tube, a very high line quality for graphic images, and large image capacity. Graphics applications are primary uses for these tubes. Both the refresh and storage tubes are very expensive and multiple colors are not yet available for them.

The third type of CRT is the raster refresh or television-type tube. It works by synthesis of a picture from a matrix of elements called pixels. A standard U. S. television has a matrix of 480 rows and 512 columns, or 245,760 pixels. The basic advantage to this type of display is the ability to have color images and up to 64 gray tones in intensity. These screens have jagged diagonal lines and poor resolution when compared to the refresh or storage tubes. Thus, the raster tube is most often used in nongraphics applications or in graphics where the lines are mostly horizontal and vertical, or where jagged diagonals are permissible. There are very high quality raster tubes, called high resolution tubes, with as many as 500,000 pixels. These high-resolution tubes produce better diagonal lines, but the main advantage to the raster display is their ability to produce multiple color images. All small commercial computers with full color graphics use raster tubes. This may change as technology improves the ability to make color storage tubes.

Typed keyboard input to the computer is frequent but it is often desirable to read data from an existing document and send it to the computer. For text, the device used to read and send the image is an Optical Character Reader (OCR). If the document is a mechanical drawing, the device used is a digitizer. The drawing is traced on the digitizer and it converts locations and lines to data bits. Input can also be from Universal Product Code (UPC) symbols and from directly monitoring electronic sensors.

Engineering computer output takes several forms but the most frequent forms are CRT displays, line printer reports, electrostatic copier, and ink-jet printer or graphics plotter drawings. Direct on-line computer access is usually via CRT with a request needed to generate hard copy such as reports or plots. Electrostatic copiers are hooked to a CRT to provide paper copies of the display image. Drawing production uses plotters or ink-jet printers. Multi-color pen plotters, available today, are increasing in popularity as report figure generators and for overhead projector slide production. Office automation using personal computers, is being increasingly used to produce and revise reports as well as to reduce design analysis and project planning time.

Computer systems are flooding every facet of engineering today. The computer is being employed to produce superior designs graphically and analytically, to optimize designs through techniques such as finite element analysis, and to simulate dynamic end use conditions. Computer systems are also being used to provide up-to-date plant layouts for plant engineering, tool designs for manufacturing engineering, and to optimize process and material flow. In the areas of analysis and planning, computers do machine justification analysis, project planning, methods and standards analysis, to mention just a few of the tasks. Unlike the slide rule, calculator, or adding machine, computer systems have the capability of producing problems as fast as they solve others. The major difficulty with computers is keeping the data they use up to date. Gallois's revelation states the difficulty well: ''If you put tomfoolery in a computer nothing comes out but tomfoolery. But this tomfoolery, having passed through a very expensive machine, is somehow ennobled and none dare criticize it.'' The use of personal computers at every work station has only made the data accuracy problem worse; every computer can now have different data.

Even in well-run companies where personal computers are the same and great care is taken to provide everyone with the same software programs, problems arise if data are entered manually from mainframe printouts. For example, one company elected to computerize all departments—in addition to their main business computer—because some applications run better and faster on small desktop machines. After installing the small personal computers in several departments, conflicting data began to surface. During one meeting to discuss the acquisition of a new manufacturing line, engineering presented its justification analysis. The controller and the sales manager both disputed the figures presented by engineering based upon their own computer analyses. However, the sales and accounting analyses didn't agree either. A week-long search ensued to determine how the numbers could be so different. All departments had the same analysis software on identical personal computers to eliminate these difficulties. The problem was uncovered. All three were using different manually entered data from the business computer system because none had direct access to the mainframe records. The problem was not in management judgment, in depreciation method or other option selections; it was in the basic data used. The real problem was eliminated by getting all the personal computers direct electronic access to current mainframe data.

In another case, engineering had just obtained a new CAD system and had completed entering its drawings on that system. The CAD system included a bill of materials software package. Part of the CAD system justification was to improve bill of materials accuracy. Engineering revised the bills using their system to make them conform to the drawings. The engineering computer could not communicate electronically with the business machine, however. Engineering had accurate bills but the business had two separate bill files to keep in step. The goal of bill accuracy was met in engineering. Bill accuracy had not improved for material control because data processing could not keep the mainframe in step. The software and CAD system was a success but the company was no better off because of system isolation and incompatibility. With two files, it is no one's fault when discrepancies surface. Engineering claims their bill is correct and they are not responsible for the mainframe data. As hard as data processing tries to get the errors fixed, engineering changes introduce new ones. The odds are that this company will never get the two files to match. The real solution is for management not to permit multiple computer installations that cannot communicate with each other. The ideal system is diagrammed in Figure 4-1.

Figure 4-1. Integrated Manufacturing Data Base

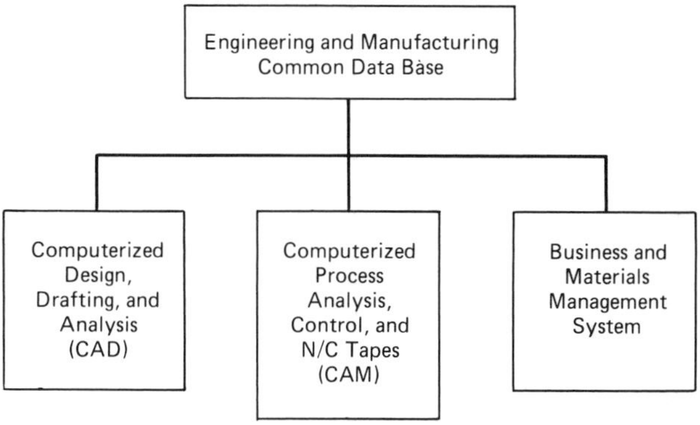

```
              ┌─────────────────────────────┐
              │  Engineering and Manufacturing│
              │      Common Data Base         │
              └─────────────────────────────┘
                            │
        ┌───────────────────┼───────────────────┐
┌───────────────┐  ┌───────────────┐  ┌───────────────┐
│ Computerized  │  │ Computerized  │  │ Business and  │
│   Design,     │  │   Process     │  │  Materials    │
│ Drafting, and │  │  Analysis,    │  │ Management    │
│   Analysis    │  │ Control, and  │  │   System      │
│    (CAD)      │  │  N/C Tapes    │  │               │
│               │  │    (CAM)      │  │               │
└───────────────┘  └───────────────┘  └───────────────┘
```

MANUFACTURING APPLICATIONS

In spite of the lack of integration of microcomputers with business computers, CAD/CAM and computer systems today are finding increasing use in industry. Here is a short compendium of applications for various industries:

Plant Engineering—for layouts, maintenance, and parts inventory

> Food and related products
> Textile mill equipment
> Tobacco products
> Lumber and wood products
> Paper and allied products
> Chemical and allied products
> Petroleum and coal
> Primary metals
> Fabricated metal products

Design Engineering—for product design, pattern layouts, and product packaging, etc.

> Electronics and electrical products
> Instrumentation and control products

Food and related products
Textile mill equipment
Apparel and allied textile products
Tobacco products
Lumber and wood products
Furniture and fixtures
Leather products
Ceramics and glass products
Paper and allied products
Chemical and allied products
Primary metals
Machinery manufacturers
Fabricated metal products

Manufacturing Engineering—NC tape generation, process design, process optimization, tool design, process automation, and special process equipment design, etc.

Electronics and electrical products
Instrumentation and control products
Food and related products
Apparel and allied textile products
Tobacco products
Lumber and wood products
Leather products
Ceramics and glass products
Paper and allied products
Printing and publishing
Chemical and allied products
Primary metals
Machinery manufacturers
Fabricated metal products

This list is by no means exhaustive but it is indicative of the pervasiveness of computer technology. Few doubt that computers and CAD/CAM systems work effectively. The big questions today are "How will they work for us?" and "What do we need to know to apply them successfully?"

To appreciate more fully the benefits of computer systems for engineering, we need to look at the list of benefits derived from these systems. The first of these and the most often cited is productivity improvement. Productivity has improved in a number of areas (a summary is shown in Figure 4-2).

Drafting.   Drawings with repetitive features are produced and updated more rapidly. Design by combination of standard features is more easily accomplished. See Figure 4-5 for typical values.

Documentation.   Bills of material and technical illustrations are more quickly produced. Technical manuals produced on word processors are made more quickly and accurately.

Design.   Calculations of stress, strain, volume, surface area, weight, thermal flux, motion, vibration, and so forth are more efficient on a computer. Assembly difficulties can be found and eliminated by assembling parts by computer. More design possibilities can be checked.

Estimating.   The ability of computers to associate data elements with text data can assist estimators in developing cost estimates for products. This benefit is even greater where integration of systems is accomplished so process as well as design data are available.

Project Analysis.   The Program Evaluation and Review Technique (PERT) and the Critical Path Method (CPM) are available today even for small desktop machines. Analysis of a project can be done by computer for projects far too complex and long for manual methods.

Scheduling.   Although scheduling is considered a part of the management function, computers also reduce the time consumed to prepare and revise engineering schedules.

Manufacturing.   When the CAD system has been successfully interfaced with numerical control tape production and program generation needs,

Figure 4-2. Productivity Improvement Summary

> Drafting
> Documentation
> Design
> Estimating
> Project analysis
> Scheduling
> Manufacturing

the computerized systems speed introduction of new and better manufactured products.

The second benefit area is better management. These benefits are fairly easy to identify but are difficult to quantify in monetary terms. A summary of the management benefits is shown in Figure 4-3.

Engineering Data Management. Computers store data in single or multiple data files which permit easy organization of data. Access to the data may be controlled by password, thus forcing some data to be on a need-to-know or change basis. Accuracy is improved by having the person responsible for developing the data enter it.

Data Distribution. Computers with communications software can communicate with other computers either in the main plant or at distant plants using telephone or satellite relay. This makes up-to-the-minute data available worldwide as needed.

Project Management. Day-by-day changes in project status can be accessed by a computer. This makes project control and status tracking easier, more accurate, and timely, and thus improves management decisions. Project-to-project consistency is dramatically improved because new project plans are developed using easily accessible data from previously completed ones.

Project Control. As projects progress and status data become available, a new analysis of the critical path can inform management where potential trouble will occur. This provides much tighter control of critical projects.

Project Scheduling. Project schedules can be revised as frequently as desired and newly discovered tasks not originally foreseen can be incorporated. This provides for improved schedules and better budgets.

Production Scheduling. Flexible scheduling of process machinery to achieve best utilization and work flow is important to manufacturing. Computers permit easy access to the data and rapid analysis to determine the best

Figure 4-3. Management Benefit Summary

Engineering data management accuracy
Data distribution speed
Project management consistency
Project control improvement
Project scheduling accuracy
Production scheduling accuracy
Budgeting and estimation accuracy

plan. The plan developed reflects reality more accurately when the data used to develop it come from up-to-date data files.

Budgeting and Estimation. When computers are used for estimating and budgeting, there is less likelihood of leaving data out or of overlooking valid costs. This improves both the completeness and the accuracy of the data.

Computer systems should be able to communicate with each other and with the main company business mainframe to prevent data consistency problems. Design, manufacturing, and plant engineering need integrated computer power. To stand alone is to stand apart. Engineering needs to be an integrated part of the manufacturing team. Most computer system manufacturers have designed the equipment to permit intercomputer communications, but many companies have not implemented the software necessary to make this a reality. The details of intercomputer communications are discussed in Chapters 5 and 6, "Computer Programming" and "CAD/CAM System Interfaces."

Benefits in the third area are virtually impossible to quantify monetarily but they can be identified. These are:

Methods Standardization. Methods used in the plant can easily be standardized using computer systems because access to methods studies is almost instantaneous. Thus operations can be standardized and methods can be improved fairly easily.

Process Standardization. Processes are combinations of individual processing sequence modules. Computer-aided process planning (CAPP) permits analysis of processes for common sequences and leads to development of standard operation sequence modules. This process standardizes and permits process quality improvements to be applied in many common locations.

Quality Drafting. Drawings produced by CAD systems are consistent in line width, numeral, and text. This produces high-quality drawings that are easy to read. This consistent quality improves user interpretation accuracy.

Improved Job Turnaround Time. Drawing production, drawing revision, and new product development times are reduced by computerized systems. This increased speed of handling routine activities reduces job turnaround times.

Improved Professional Development. Professional development is enhanced when routine tasks are computerized. The time released can be applied to professional development. Additional education in material control, new modeling techniques, process planning, manufacturing plant operations, and management development is needed in plants using integrated systems.

Professional development in these areas pays handsomely in improved communications and understanding plantwide.

Improved Morale. Morale improves as routine tasks consume a smaller fraction of an employee's time. Automation of routine tasks and viewing employee development as a priority item result in improved job satisfaction and higher morale.

A typical CAD/CAM system configuration is shown in Figure 4-4. Graphics systems are little different from business computers. The primary long-term storage medium is magnetic tape. This is both the most convenient way to store vast quantities of engineering data and a fairly quick way to retrieve them. Magnetic tape, however, is a sequential filing system. Data are stored end to end along the tape, and data located near the far end of the tape take several minutes to retrieve

By contrast, the hard disk drive is a rapid access medium. This is a spinning metal disk with a magnetic coating. This disk is formatted into cylinders

Figure 4-4. Typical CAD/CAM System Configuration

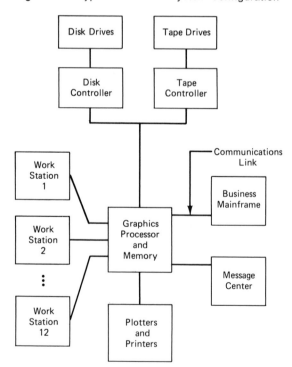

called tracks and subdivided into angular sectors. Both data and a disk directory are written to the disk. This directory is an index of the track and sector location of each data file. The directory permits the computer to locate the data by looking up the file coordinates and going directly to that position. An analogous procedure would be using an index to look up a topic in a book. Unlike the tape, however, disk access is extremely rapid because it is not sequential.

Both the tape drive and the disk drive need a controller. The controller is an electronic circuit that converts the message to find a file with a certain name into series of commands that allow the tape drive or disk read heads to locate the file. Controllers also determine the bit density per media inch, where the directory is, and the order in which bits are written to the disk or tape.

Work stations are CRTs, digitizers, light pens, graphic tablets, and other human interface tools. The plotters and printers provide hard paper or mylar film copies of graphics and/or text output.

The communications link to the business mainframe permits the CAD system to transmit and receive information located in the business data base.

CAD/CAM PRODUCTIVITY MEASUREMENT

To measure CAD system productivity correctly, the design and drafting process must be thoroughly understood. Reference activities (looking things up in catalogs or manuals, for example) consume much design time. These activities are common to both CAD and non-CAD operations. Productivity computations require removal of common activities. Computerized tables reduce reference time. It is necessary to include this reference time in calculations, however. One of the easiest ways to determine the CAD and non-CAD operations is to do a work sampling in the two environments. Many CAD system suppliers have customers who will permit a work sampling in their facilities for "after the installation"data. Estimates for non-CAD work and samples of the common activities provide data even if samples cannot be obtained from both environments.

CAD system productivity is the ratio of the hours expended before CAD to the hours after. If total design hours before CAD are given by HB, reference hours by HR, and total design hours after CAD by HC, then:

$$\text{Productivity} = \frac{\text{HB} - \text{HR}}{\text{HC} - \text{HR}}$$

If we define MR as the average manpower cost rate and CR as the CAD system console cost rate ($/hour), the productivity value is a gross estimate of the maximum console cost rate to break even with a manual operation. This is because:

$$\text{Productivity} = \frac{HB - HR}{HC - HR} > = \frac{MR + CR}{MR}$$

Thus

$$CR < = \frac{MR (HB - HR)}{HC - HR} - MR$$

Experience has shown that:

$$\frac{MR (HB - HR)}{CR} = \text{approx. } .50$$

This is based on operations running two shifts with 80 percent of the CAD available hours used. The typical productivity is approximately 2.7 to 1.

CAD/CAM PRODUCTIVITY RATIOS

There is much discussion and controversy about expected productivity ratios when computer systems are instituted. Some CAD equipment suppliers quote ratios as high as 7.0 to 1 for their systems. Although productivity ratios can reach those quoted by system manufacturers, Figure 4-5 shows some actually measured performances by type of drawing being produced. These are representative of those users can expect. Writing automatic menu-driven design programs improves productivity dramatically. Design programs that use basic standard product line size and produce completed drawings develop productivity ratios in the 5.0 to 10.0 range.

Figure 4-5. Typical Actual CAD/CAM Productivity Ratios

Design by menu selection	7.0
Simple logic drawings	5.0
Single-line drawings	4.0
Wiring diagrams	3.5
Piping and instrumentation diagrams	3.5
Simple product drawings	4.3
Assembly and detail drawings	3.7
Sheet metal drawings	3.7
Extrusion drawings	3.2
Numerical control tapes	2.7
Detail new product drawings	2.4
Tool design drawings	2.1
Structural steel drawings	1.7
Piping layout drawings	1.5

CAD/CAM SYSTEM COSTS

System costs vary widely with the advent of the personal computer. Many 16-bit personal computers support simple CAD wire frame drawing software. These systems can do plant layout, simple product drawings, and assembly drawings, but they are not intended for fine detailed work nor for complex drawings. They do give an engineer the capability to make sketches and to do designs in the 8½ x 11 to 11 x 17 size with a small commercial plotter. The cost of a complete personal computer with graphics software and plotter is between $4,000 and $10,000. These systems are true graphics CAD/CAM systems and they are a viable alternative for small businesses whose drawing needs are simple and limited.

CAD/CAM systems usually cost between $150,000 and $500,000. A typical single-terminal system will cost about $100,000 for hardware and $50,000 for software. This can vary upward to $300,000 for some single-user systems depending on the hardware and software purchased. A typical four-terminal system starts at around $250,000 and goes up to about $500,000. Systems can run over $500,000 but these systems are not usually dedicated to the CAD/CAM functions. The parameters which affect system cost are shown in Figure 4-6.

The following is an item-by-item explanation of the factors of system cost:

Hardware. Consists of the Central Processor (CPU), the graphics display terminals or CRTs, and digitizers to take existing drawings and convert the data to electronic images. Printers print text and obtain fast small prints, plotters make the drawings, and disk and tape storage systems retain them.

Software. Consists of the computer operating system which polices the movement of data to the peripheral devices, a database manager which stores and retrieves the data from disk and tape files, and functional software used to do specific tasks such as structural analysis, control circuit design,

Figure 4-6. Factors Effecting System Cost

Hardware
Software
installation cost
System maintenance cost
Site preparation cost
System operation cost
Depreciation
Interest rate on the investment
System downtime cost
Utilization

vibrational analysis, heat transfer modeling, mathematical analysis, and optimization.

Installation.   Is the cost of physically installing the hardware and software in the facility.

Maintenance.   Is the cost of providing preventive and breakdown maintenance for the hardware and software. Software maintenance also involves identifying and correcting errors.

Site Preparation.   These are the costs required to prepare the area for the hardware. They include walls, floors, power, fire protection, other utilities, and work station furniture and supplies.

Operation.   These are the costs of administering the system. They include:

Work Accounting: Receiving, log-in and -out, work-in-process tracking, job priority, plotter scheduling, and so forth.
File Maintenance System: Backup copy system; drawing filing system; archival storage for tapes, disks, and drawings; file indexing systems; secure storage, and so on.
Training System: Personnel development, CAD usage training, remedial training, apprenticeship programs, and the like.

Depreciation.   The cost associated with the decrease in utility of the hardware and software with time.

Interest.   The cost of borrowing the funds to purchase the system.

Downtime.   The cost of having the equipment inoperative includes such items as overtime to recover, late project costs, and production delay costs, etc.

Utilization.   Costs for utilization are overheads normally applied to equipment. These are inversely proportional to utilization. The less the equipment is used, the higher the overhead applied and vice versa.

The financial computations for justification of equipment, and specifically computer equipment, vary considerably among companies. Actual financial analysis of these items is left to the reader. There are, however, several intangible benefits to consider in a financial analysis. Improved project scheduling, control, and budgeting, as well as standardization of products and processes are very real and tangible assets. Just because they are hard to quantify, don't ignore them. Better flow in the plant will result from these benefits.

Computer systems represent the hardware side of the integration of engineering. The software that works within the hardware is essential for true integration of functions within a manufacturing environment. Chapter 5,"Computer Programming," lays the foundation for understanding the evolution of software and the vital role it plays in systems.

5

COMPUTER PROGRAMMING

The advent of the small hand-held programmable calculator rendered the slide rule obsolete. This doesn't mean that engineers don't use slide rules anymore, but slide rules are horse and buggy antiques in the computer age. The programmable calculator does complex mathematical analysis much faster than a slide rule. The 8- to 15-digit precision of calculators have provided numerical results that were previously too imprecise using the three significant digits on a slide rule. This power has improved engineering productivity. Some people believe that this is only the first step in a vast improvement possible with computers. To realize this potential requires a detailed understanding of the management of the computer resource. The previous chapter provided the hardware foundation; this chapter covers the programs or instruction sets that make the hardware useful.

Computer literacy is a must for engineering and management, as well as many other fields today. In fact, most engineering schools require a foundation in computer programming to obtain a degree. Engineers and managers must understand the pros and cons of having various programming languages, including the difficulties caused by having many languages and the obstacles multiple languages present to better integration. This chapter gives the non-computer professional a foundation in application languages for computers and provides a basis for further study of them.

PROGRAMMING LANGUAGE CONCEPTS

Keying data and operations into calculators yields numerical results. These keystroke sequences are really small programs. Programmable calculators are programmed using instructions (keystroke sequence abbreviations) very similar to those of computer assembly language. A computer program, at its simplest level, is the set of instructions which causes a computer to follow a planned sequence of steps or operations. The sequence directs, controls, or channels data in accordance with a plan, schedule, or code. If we translate this definition into the simplest of examples, the computer works using the following cycle:

1. The control unit retrieves an instruction.
2. The instruction is converted to an electronic signal.
3. The arithmetic, memory, or peripheral device carries out the instruction.
4. The cycle repeats beginning at step 1.

A program consists of many instructions executed at very high speed by the processor. Computer programs are not as simple as the above four-step sequence indicates, however. The instructions must be ordered in such a way as to produce the desired result. Thus, there are two aspects to programming: the instructions needed and the order of those instructions. Every program is a network of instructions that define the task the computer is to perform, so that programming is really the creation of instruction sequences.

BINARY MACHINE LANGUAGE

Any discussion of computer language must begin with machine language because it is the language understood by the computer. As background information, it is useful to have some conceptual understanding of machine language programming and the reason few people write programs in it. The following series of instructions in machine language performs addition of two numbers:

$$00111010$$
$$01100000$$
$$00000000$$
$$01000111$$
$$00111010$$
$$01100001$$
$$00000000$$

10000000
00110010
01100010
00000000

This series of binary numbers, consisting of line-by-line instructions to the processor, results in the addition of two numbers on a Z-80 processor. That is not obvious to most people. The simple switching of a single digit from a 1 to a 0 or vice versa changes the program. Computers follow instructions exactly and thus programs cannot be 99.999 percent accurate—they must be 100 percent. This makes for difficulty and high error rates when it is necessary to program in machine language.

HEXADECIMAL MACHINE LANGUAGE

Fortunately, someone once noticed that the binary 8-bit code (called a byte) could be grouped in two 4-bit groups (called nibbles). That person converted each 4-bit group to a base 16 number, which reduced programming errors.

Conversion of the 4-bit binary to base 16 or hexadecimal proceeds as any other number base conversion. The hexadecimal number set is just like our normal base 10 system with six more digits.

The base 10 number set is:

0   1   2   3   4   5   6   7   8   9

The base 16 (hexadecimal) number set is:

0   1   2   3   4   5   6   7   8   9   A   B   C   D   E   F

Changing number bases may be difficult and confusing but for a machine language computer programmer it was an error reducer. The addition program in the previous example now becomes:

Convert Binary to Hexadecimal		Program in Hexadecimal
0011 = 3	1010 = A	3A
0110 = 6	0000 = 0	60
0000 = 0	0000 = 0	00
0100 = 4	0111 = 7	47
0011 = 3	1010 = A	3A
0110 = 6	0001 = 1	61
0000 = 0	0000 = 0	00
1000 = 8	0000 = 0	80
0011 = 3	0010 = 2	32
0110 = 6	0010 = 2	62
0000 = 0	0000 = 0	00

Eventually the program written in hexadecimal will require translation to binary for use by the computer. This process is automated by a programmed read-only memory (PROM) chip containing a program to do the conversion. Programming thus moved one step from machine binary code to hexadecimal code. Hexadecimal code increased program accuracy because it decreased the number of digits needed and increased the number of character differences. However, the program was still not recognizable as one that performs addition.

ASSEMBLY LANGUAGE

Recognizability was improved by converting the hexadecimal codes to the mnemonic codes of assembly language. Programs contain two element types: operation codes and memory locations. Operation codes are instructions to the processor to perform a task, while memory addresses store the data in the computer's memory. Processors use an operation code to signal that the two hexadecimal numbers immediately following define a data memory address. The code 3A in the preceding program is that code. Whenever the processor gets a 3A, it uses the next two bytes as a memory address. The addition program can be deciphered into the following:

3A—Load the following address into the accumulator

60⎫
00⎭ The memory location 6000 contains the data

47—Move data from accumulator to register B

3A—Load the following address into the accumulator

61⎫
00⎭ The memory location 6100 contains the data

80—Add register B to the accumulator

32—Load the following register with the result

62⎫
00⎭ The memory location 6200 contains the data

	Hexadecimal Program	Assembly Language Mnemonics
	3A ⎫	
	60 ⎬	LD A,(60H)
	00 ⎭	

Hexadecimal Program	Assembly Language Mnemonics
47	MOV B,A
3A 61 00	LD A,(61H)
80	ADD A,B
32 62 00	LD (62H),A

This assembly language program is still not understandable to the average nonprogrammer, but at least it has some instructions that resemble abbreviations. LD resembles load, MOV looks like move, and ADD reads directly into English. These abbreviations or mnemonics were so much better than 3A, 47, and 80 in hexadecimal or the binary equivalent that programmers saw this as a gigantic leap forward. Programs were much easier to write and more error free when written in assembly language.

As mentioned earlier, programmable calculators program in a language that looks surprisingly like computer assembly language. The preceding program written for a Hewlett Packard programmable calculator would be as follows:

RCL 2—Recall the value in Register 2
RCL 3—Recall the value in Register 3
+   —Add the values
STO 4—Store the answer in Register 4

Programmable calculators are really small-scale assembly language, number processing computers. It is also useful to note that command programs (called macros) for CAD/CAM systems are essentially assembly language instruction sequences.

To get from assembly language back to machine language, which the computer requires, a conversion must be done. This conversion process is called assembling or compiling. A computer program called an assembler or compiler does this translation. The compiler's sole function is to take a program written in some other language and convert it to machine binary code for execution. In a real sense, the compiler or assembler is a language translator. It translates from some human understandable language to a machine understandable one. To perform this task, the translator requires the human understandable language to match in form (called syntax) and in words or mnemonics the language the assembler or compiler uses.

Assembly language is the most fundamental and flexible programming language. Anything can be done in assembly language, and many consider it to be the best computer language because of that flexibility. Assembly language is also called the worst language because it is so complex and machine dependent that programs are composed of pages of statements to accomplish relatively simple tasks. Machine dependency results because assembly language uses direct processor commands, and these commands are not the same for all processors. Despite the pros and cons of assembly language, it is very useful for certain tasks.

Today most assembly language programs do process control or language compilation. These include language compilers, HVAC monitoring and control, traffic light control, computer operating systems programs, missile and gun control, and pollution system control. Robots are programmed using assembly language, as are short routines in computer programs such as searching a file, sorting a file, adding a record to a file, or performing a counting of records. The hallmarks of well written assembly language programs are speed and efficiency; that is, they usually run the fastest and consume the least memory.

HIGHER LEVEL COMPUTER LANGUAGES

Not long after assembly language was developed, programmers sought to simplify their lives further. What they wanted was a language that would take the assembly program on pp. 63–64 and reduce it to C = A + B. They also needed the ability to write a program and run it on different machines. Assembly language is machine dependent. Each particular processor has its own assembly instruction set, and different processors have different instruction sets. Assembly language programs were not transportable from machine to machine. Getting an assembly language program from one machine to another requires a complete rewrite of the program.

These needs for a machine-independent language were filled in the mid-1950s. A language called FORTRAN was developed. FORTRAN is really an acronym for FORmula TRANslation. This language was designed for scientific and technical applications where mathematical formulas were extensive. All high-level languages are a series of commands associated with assembly language routines. For example, every time C = A + B appeared in the high-level language, an assembly language routine was substituted. Now at last it was relatively easy, in theory, to write complex programs that were error free and transportable. All one needed was a FORTRAN program and a FORTRAN compiler for the computer. If someone was a programmer, FORTRAN was relatively easy to understand and rapidly became the most common high-level language of its time.

A high-level language is a procedure-oriented rather than instruction-oriented language. The programmer defines the procedures required to produce the desired result, and the compiler provides the detailed instructions in machine language.

There are two major advantages to high-level languages: Programs are much faster to write and they are easier to understand. It is common for programmers to write programs ten times faster in a high-level language than in assembly language. It is also much easier to read C = A + B than the equivalent assembly language. Programmers don't need any knowledge of computer hardware to write FORTRAN programs because the compiler takes care of those details.

FORTRAN PROGRAM LANGUAGE

FORTRAN has evolved over many years. It has been enhanced and made more useful but certain tasks and applications are still not efficient. FORTRAN was not as simple to learn and to use as some thought a language should be. A major problem for FORTRAN was its design. It was designed to do complex mathematics, but business needed primarily the simple arithmetic operations of addition, subtraction, multiplication, and division. Business handles massive amounts of alphanumeric or character-based data, and FORTRAN did not contain some of the procedures to do business applications easily. FORTRAN for business applications is similar to using a rowboat oar to paddle a canoe; it works but a paddle is better.

COBOL PROGRAM LANGUAGE

The Common Business Oriented Language (COBOL) was developed specifically to meet business needs. COBOL has very limited mathematical capability but has some highly efficient routines to handle character data and data files. With the computerization of business applications, two systems were operating side by side—in engineering, FORTRAN, and in accounting, COBOL. This may have seemed like the perfect solution, but it was not long before everyone was unhappy because the computer tower of Babel was under construction.

PROGRAM LANGUAGE COMPATIBILITY

Since COBOL and FORTRAN programs do not use the same file structures, they are unable to use the same files. This was the real beginning of the widening gap that has resulted in engineering being divorced from manufacturing and the operation of the business it serves. Accounting paddled their canoe and engineering rowed away in their rowboat, each in different directions. The problem created by two specialized languages was how to manage to get them to share their data. With specialists and separate departments, the problem of getting them both going in the same direction became monumental.

This gap grew wider with the development of the language BASIC at Dartmouth College by John Kemeny and Thomas Kurtz. Their goal was the creation of a beginners' programming language for persons from diverse educational backgrounds. The new high-level language they developed as part of a computer project for General Electric in the 1960s was called Beginners All-purpose Symbolic Instruction Code, or BASIC. It was indeed easy to learn and it has proven itself a good general-purpose language. It was not widely used, however, until the 1970s when microcomputers became more widely available. The advent of the home/business small microcomputer has made BASIC the most common of all languages. At the same time, different dialects of BASIC greatly complicated the transport of programs from machine to machine. Remember, one of the goals of high-level language was to make programs transportable. The transportability goal disintegrated, not because programmers wanted it that way but because software and hardware manufacturers wanted it that way.

Marketing considerations of both hardware and software suppliers made a patchwork quilt out of the common language goal. FORTRAN and COBOL are no different. The lack of transportability is due, at least in part, to a lack of common language standards. Nevertheless, the lack of standards played directly into the hands of marketing at the large hardware and software companies.

Marketing logic is a major reason for lack of program transportability and machine integration. Hardware manufacturers want buyers to use their equipment and the programs they can supply for that equipment. Language suppliers, of course, want purchasers to use the software on the machine available, but they also want programs to be unique to their own compiler. Thus, IBM wants only its equipment and software used, so programs written in IBM FORTRAN cannot be run on someone else's compiler. An IBM compiler will not work on non-IBM equipment so the buyer must use IBM. This is fine for IBM, Burroughs, Sperry-Univac, Digital Equipment, and the myriad of others that do likewise. But programmers and businesses dislike this situation because they have enormous difficulty in moving programs between machines. Things are improving, however; standards are being written, and languages are beginning to address that transportability problem. BASIC is being standardized, but the standards for BASIC still have a long way to go before programs will be portable.

PASCAL PROGRAM LANGUAGE

In 1971 another new language, called PASCAL, was invented. Professor Nicklaus Wirth developed it with the aim of teaching programming as a systematic discipline. Although not originally intended as a major programming language, it caught on rapidly because it had two vital characteristics: (1) PAS-

CAL was a structured language with rigid, consistent syntax rules, which forced a methodical approach to programming; (2) it handled pointers to data elements efficiently. A pointer is the memory address location for a variable.

BASIC contains an undisciplined "go to" construction which directs the computer to branch to previous or subsequent program instructions. PASCAL, by contrast, consists of in-line procedures which begin and end only in series. Sequencing these procedures achieves the program result. Goto's are not permitted if they cause jumps to instructions in structured statements. The result is modular programming, which consists of an assembly of code modules (procedures) to make a complete program.

The second major attribute of PASCAL is that it is well suited to doing graphics. Graphics requires that data elements be in sequence. PASCAL permits one data element to contain a pointer to the next element in the list. Graphics applications are a major reason for the success of PASCAL because PASCAL supports the use of pointers and linked lists better than other languages. PASCAL is also considered to be one of the more transportable languages because it is new and completely standardized. Given time, it may be enhanced into untransportability but at present programs written in PASCAL move between machines fairly easily.

C PROGRAM LANGUAGE

One more language will be addressed here—C language. C is a concise language which has a small core of approximately 30 reserved words called a Kernal. Input/Output is gathered into a library of specific but standard routines. This means that C operates easily on new processors. The language was invented at Bell Laboratories by Dennis Ritchie in 1972. The UNIX operating system, also developed by Bell Laboratories, is a large C language program. The UNIX system is rapidly becoming a standard operating system for mainframe and microcomputers because of language transportability. To transport the UNIX operating system to a different processor only requires writing a C compiler for that processor. Systems programmers say this is a relatively easy task. UNIX itself is the largest project to date in the C language since it comprises over 300,000 lines of code. The C language is available for almost every computer from 8 bit to the CRAY 1 and is thus the most transportable language at this time. It is both as powerful as assembly language and as diverse as many high-level languages. It is new, however, and many old programs and operating systems have not been converted to C. This is changing, but at present C is still a young language and application programs are still limited.

A number of other languages exist for computers such as APL, RPG, ADA, LISP, FORTH, SMALLTALK, LOGO, and many, many more. These each have their own special attributes and therefore their own best uses. We have a computer tower of Babel with the multitude of languages already in use, and the situation isn't improving much.

THE PITFALLS OF SUBOPTIMIZATION

The following scenario is so common it is tragic. Company A has a large mainframe computer to run the business but it doesn't have up-to-date hardware and software. Engineering cannot use it because of insufficient capacity coupled with the lack of a FORTRAN compiler. The solution decided upon is to purchase a modern CAD/CAM system for engineering that has the needed software.

This gets engineering what it needs without sacrificing or inconveniencing the business. A year or two later, manufacturing and accounting find they need more computer capacity and the load could be handled by time sharing with the underutilized engineering machine. Engineering's computer, however, is not able to run the business programs because the program languages and file structures are incompatible.

Now business decides to upgrade its hardware and elects to change hardware manufacturers and software languages to keep pace with the state of the art. This means that all the business application programs must be rewritten. In all likelihood, no one has considered machine integration, and the new machine will be from a different manufacturer. The company has repeated a mistake it made a few years earlier when it purchased new engineering equipment.

Many companies have attacked the problem by centralized data processing, that is, by requiring all computer equipment purchases to be okayed by the central data processing group. However, this leads to "the bigger, the better" syndrome. Central data processing wants everything done on the mainframe, and new programs and uses require justifying a bigger central computer.

However, many executives and managers continue to proliferate the desktop, microcomputer approach. These executives see no need to access the mainframe for data and continue to purchase their own hardware. The inability to share data is similar to coaching a football team when the coach and the players can't communicate with each other. Good management teams have good communications and teamwork, not independent but efficient all-stars.

This is a symptom of a disease that might be called the suboptimization syndrome. This syndrome results from the belief that the greatest productivity and efficiency come from optimizing each task independently. Unfortunately, tasks are not independent in team sports or in business, and optimization must be achieved by better integration rather than by suboptimization. For example, why perform a methods and time standards analysis to optimize a task that should not be done at all? In fact, engineers do just that when they set standards on 100 percent inspection of parts because they have not concentrated on eliminating the variations in the process to make it produce consistent high quality.

New high-level programming languages are being written daily, but the name of the game is not to have the latest and greatest, but the most compatible. Programs should allow computers in the plant to share resource data and should also be standardized within as few languages as possible. This eliminates the need to write multitudes of translator programs to permit data sharing.

COMMERCIAL SOFTWARE

Commercial software is available today to do most tasks on most machines, and many equipment and software suppliers are sensitive to this need for communications between machines. The great emphasis today is on distributed processing, which will be discussed in Chapter 6, "CAD/CAM System Interfaces." There is also a greater emphasis in conforming the business to the commercially available software rather than attempting to get software to fit the business exactly. Of course, standardization has its limits, but many commercial software packages have sufficient flexibility to accommodate most of the business variations.

There are three main reasons to use commercial software: cost, timely availability, and support. The cost of developing software is very high. It takes a long time to develop, debug, and thoroughly test the many routines that comprise a single program. Commercial software has the best test base that exists—its many users. Software companies can afford the expert programmers needed to develop and support software in the field. Thus, specialization pays in terms of product development, cost to consumers, and the expertise necessary to keep the product functioning.

Since commercial software sells essentially off the shelf, with perhaps limited modification for a specific application, the product is running almost immediately. For many companies this is important because they cannot afford to spend years getting a program running. Since time is money to most companies, the faster the product can be put on line, the less costly the product will be. This is why having tested software in other users' hands can be a significant advantage both to new consumers and to the software developer.

Software and hardware user groups are available for many widely used pieces of equipment and software packages. These groups have often discovered both the bug and what to do to avoid it ahead of the vendor. They can be very helpful to those using or considering acquisition of software and computer products. Before buying a system, ask for the name and address of the local user group; it is usually available from the hardware and/or software vendor.

CUSTOM PROGRAMMING

Custom programs are usually of three types: contracted programs, programs by outside programmers, and in-house programs. Contracted programs are customized by commercial software suppliers for a specific application, and are therefore much higher in cost than their commercial equivalent. Although a custom contract delivers the software as outlined in the specifications, the specifications must insure the programs will perform properly. One of the more frequently sought custom packages is order entry. For many companies, the order entry functions are similar but sufficiently unique that custom software is the solution.

Many companies supply programmers to do custom program work. The rates for these programmers are as varied as the companies themselves. Some bill by the hour, some by the week, some by the month, some by yearly contract, and others by the job. These companies will usually supply competent people with experience on both the client's hardware and software. The work may be done at the client's plant or at the vendor's facilities. As with a contracted program, a great deal of time is spent on program specifications. The specifications must define exactly what the program must do, what display or CRT screens are to look like, what reports will be developed, what files will be used, and so forth. The time spent to detail these specifications is time saved during programming, and the up-front investment returns itself many times over in avoiding rewriting programs that don't quite do the job.

IN-HOUSE PROGRAMS

A third method of obtaining custom programs is to write them in-house. For some programs, in some applications, this is cost effective and desirable. The major advantage of in-house programming is that company people understand company problems better than outsiders, so the learning curve time needed to teach outsiders the business drops. A second advantage is that this same inside knowledge makes for ease of modification when needed. The major disadvantages to in-house programs are their unique nature. Many times the flexibility to handle business changes is not built into the programs. This, on occasion, forces total rewrite of programs.

On CAD/CAM systems, the "macro" routines are best written by in-house personnel. Major utilities are best left to commercial software, and integration or networking is best done with custom software if commercial is not available. Perhaps the best path is defined by the following guidelines:

1. Buy commercial software and make every effort to use it as provided. Try to fit commercial software into the business. This does not mean allowing the software to dictate how the business will run, but it does mean that standard software provides the data most companies use. If most companies in your industry can use commercial software, find out why you are different. Being different is not necessarily desirable.

2. Buy custom software only for applications for which commercial varieties do not exist or when the company's needs are so different that the off-the-shelf variety won't work.

3. Write only those programs in-house that are short, unique routines with well-defined needs and where company-specific knowledge is essential. On CAD/CAM systems, design macro routines developed by engineering, keep design information classified, and provide programs to enhance productivity.

DESIGN BY COMPUTER

Design by computer takes many forms. One form is to use the computer as a number and data processor to analyze information based upon a set of predefined criteria. The program produces a report on a CRT or printer that provides data to draft an on-paper design. This approach is the design-utility approach because the computer may do more than calculate but in essence simply compiles data used to produce a drawing by manual techniques.

The second approach is the semi-automated design approach. This approach is similar in that data are supplied by users, computations are done, and output is printed. The output does not consist simply of data but data in the form of dimensions located so drawing overlays produce finished drawings.

The third approach is fully automated design. The input data are used to drive a plotter which then produces a finished drawing complete with all dimensions to scale.

The design-utility approach has been successfully used for many years by many companies. This approach has produced reams of reports and is by far the most common approach to design by computer. It provides the greatest flexibility and the least custom programming. Commercial software is available to do many types of analysis, including finite element modeling, vibration analysis, electronic circuit design, and building ventilation design. More programs are becoming available that extend this capability. In addition, graphics software is not required for this technique, and many of these analysis programs run on desktop machines. The use of utility programs to provide the necessary analysis data is growing so fast today it is hard to keep track of what is available. The equipment manufacturer and user groups are the best source of information on available utility software.

The semi-automated design approach was used very successfully by many companies to automate product design for over 15 years. The drawings were produced on clear mylar, and the computer placed the dimensions where they were to show up. Since the drawings did not have to be to scale, this approach worked very well. For many simple products whose design is defined by a set of equations for which input data are known, this approach is quite cost effective. It is also applicable to many desktop computers.

The fully automated design approach is still being pioneered by companies with CAD/CAM systems. Simple designs, complete with scale drawings, have been produced by several companies. The aircraft industry has been producing drawings of parts this way for many years. The automotive industry is doing it on an increasing number of parts. It may be a long time before this becomes common practice, however, because many products, tooling, packages, and machines don't lend themselves to mathematical design. The most common technique for "nonstandard" designs will remain the utility-design approach. The day is approaching when computers will be able to design automatically and produce drawings of standard parts for a majority of components used in assemblies. This is truly a new frontier.

STANDARDIZED SOFTWARE FOR MANUFACTURING CONTROL (MRP II)

Consultants have defined and published standard software for manufacturing control (MRP II) systems. Specific characteristics were incorporated into that definition which made it a standard for batch manufacturing. However, the repetitive and just-in-time environments did not function the same way as batch. The standard software described a transaction-based system with the following characteristics:

Orders are issued for discrete quantities.

Individual lots or orders are identified.

Components and assemblies are issued to orders.

Labor reports to operations on the order.

Finished orders increase inventories.

Costs accumulate to the order.

The above repeats at all BOM levels.

This results in the following data flows for material in the standard system:

Orders are issued to vendors for a fixed component quantity.
The vendor ships the quantity ordered.
Receiving issues a receipt transaction when the parts arrive.
The receipt transaction increases component inventory.
→Shop production orders are issued to manufacturing.
R Materials requisitions issue components to the shop.
E Material requisitions are:
P   Used to obtain parts from warehouse.
E   Used to decrease component inventories.
A Manufacturing produces the parts.
T Parts are passed to stock location.
S Shop order closes increasing part inventory.
→Next BOM level occurs until final assembly is complete.
Final assembly is received by finished goods warehouse.
Finished goods inventory increases.
Shipping fills customer orders from stock.
Shipping order closes to decrease finished goods inventory.

This system assumes discrete transactions at each point in the process. This standard system must be modified for repetitive or just-in-time environments because the environment functions differently. The repetitive environment characteristics are:

Production rates per period are used.
No specific lots exist.
Materials are issued to work in process (WIP).
Labor reports to the process.
Period completions backflush WIP inventory.
Costs accumulate to the process.
All processes are part of end item.

This is a two-transaction, not a many-transaction system. Materials are issued to WIP to permit production to occur. Completed products and the bill of materials for the item reduce the WIP inventory. Individual work orders do not really exist as they do in batch manufacturing. The repetitive system materials data flow is as follows:

Order issued to vendor for a fixed component quantity.

Vendor ships quantity ordered.

Receiving issues receipt transaction when parts arrive.

Receipt transaction increases WIP inventory.

Manufacturing produces finished goods.

Completed products are received into finished goods:

This depletes the WIP by backflush.

This increases finished product inventory.

Shipping fills customer orders from stock.

Shipping order closes to decrease finished goods inventory.

This repetitive manufacturing environment is rate controlled rather than batch controlled and has fewer transactions than the batch environment. It also has fewer points where processing stops long enough to permit easy data gathering. For many companies, attempting to implement just-in-time techniques may require modification of the software for manufacturing control. Careful choice of software can provide a standard software package that will operate in both of these environments in the same plant. This is a real plus since some manufacturing areas may lend themselves to a repetitive environment while others within the same plant are batch in nature.

The defense and aerospace industries need software which is different from either of the above two. They need software that handles product costing by contract and/or progress payments and inventory control by project. Software must address the intended environment and the "standard software" definition applies only in a limited batch production environment. Therefore, management must exercise care to choose software that fits the environment and not simply settle for one that some consulting firm has deemed "standard."

6

COMPUTER SYSTEM INTERFACES

No discussion of computerization in the manufacturing firm is complete without some mention of computer communications, known as *interfacing*—a field in its own right. Computers acquire data from many sources, among them, keyboards, card readers, light pens, digitizers, joysticks, mice, trackballs, tape drives, and disk drives. These devices primarily transfer data from people to the computer. Computers also acquire data from process controllers, sensors, measurement instruments, and other computers, to name a few. Interfacing is the field covering communication among computers and other devices.

Parallel and serial interfaces are the two methods of transferring data from another device to a computer or between computers. The oldest interface is serial and employs the human being. This is the operator interface. An operator looks at a panel of meters or printout and types the data into a card punch or CRT keyboard. This method is slow, inaccurate, and crude. The rate at which the operator performs a complete data entry cycle is much too slow for modern test systems or industrial sensors. Data exchange between computers using the operator interface became obsolete in the early 1960s.

The first efficient communications method is known as a parallel interface or bus. The parallel interface bus uses one wire for each data bit. At least eight wires exist to transfer data on an 8-bit system. In practice, it takes as many as 36 wires to provide parallel communication. All internal data transfers—that is, within the computer processor—happen in parallel. The advantage to this is that data transmission occurs as fast as the processor can send and receive data. The transmission speed is the internal clock rate of the computer. Clock rates vary from four million bits per second for an 8-bit microcomputer to around two million, million bits per second for supercomputers.

PARALLEL COMMUNICATIONS

Parallel interfaces between microcomputers and sensors or process control devices are usually direct bus plug-in. The interface is a printed circuit card plugged directly into the computer's busway. Wires connect the card to the external device or process controller. Interface cards for hard disk drives, floppy disk drives, color monitors, and printers are built into modern microcomputers. The interface card contains the hardware and/or software necessary to connect the sensor or controller to the computer.

Parallel interfaces among the 8-bit and the 16-bit personal computers and peripheral devices are standardized. The interface standard covers the wires required in the parallel interface but ignores the order of the connector pins. The hardware manuals supplied with computer systems and the peripheral device manuals tell which pins to use for particular signals. Most parallel devices use a 36-pin Amphenol 57-30360 connector. The connector on an Altos 8000 series computer uses 16 pins for signals and three pins for power; the remainder are grounded, float, or unassigned. The Altos uses the pin connections shown in Figure 6-1.

The pins noted as having signals going in or out are the vital few required by most peripheral devices. It takes 16 signals plus a ground to satisfy most devices for parallel communications. It is essential that peripheral devices receive and transmit to the computer over the correct wires. Standard cables are available for most common computers and peripheral devices. Most personal computers do not communicate over parallel cables to mainframe computers. They use a serial interface instead. Peripheral devices such as printers commonly use parallel interfaces for communication with personal computers.

SERIAL COMMUNICATIONS

The second method of interfacing efficiently is serially. The principal difference between serial and parallel communications is that a single wire carries serial data for exchange. Telephone lines or microwave links for long distance communication use this method. The wiring is much simpler than the parallel. Figure 6-2 shows the pin connections for the standard RS 232 serial connector.

This connector is so common and standardized that almost all equipment uses the same configuration. The transmit and receive data pins (2 and 3) cross over between the sending and receiving units as do pins 4 and 5. This lets the computer send data out pin 3 and the receiver send data back using pin 3 at that end. The cable swaps pins so pin 3 at one end is pin 2 at the other.

Now comes the complexity. There are two types of serial communication—synchronous and asynchronous. It is considerably easier to understand serial communications if these types are treated separately. The principal characteristic of synchronous transfer is that data are in a continuous stream be-

Figure 6-1 Altos 8000 series Parallel Port
Pin Assignment

Pin No.	Use	Data Direction
1	Data Strobe	Out
2	Data 0	Out
3	Data 3	Out
4	Data 1	Out
5	Data 6	Out
6	Data 7	Out
7	Acknowledge	In
8	Busy	In
9	Data 2	Out
10	Data 4	Out
11	Data 5	Out
12	Control	Out
13	Select	In
14	+5 volts	—
15	Paper Empty	In
16	−12 volts	—
17	Input Prime	In
18	Floating	—
19	Floating	—
20	Ground	—
21	Ground	—
22	Ground	—
23	Ground	—
24	Ground	—
25	Ground	—
26	Unassigned	—
27	Ground	—
28	Ground	—
29	Ground	—
30	Ground	—
31	Ground	—
32	Ground	—
33	Ground	—
34	Fault	In
35	Unassigned	—
36	+12 volts	—

tween a synchronized sending and receiving unit at the sending and receiving computers.

The receiver and transmitter are synchronized during data exchange to define data unit boundaries. Each data unit (usually one byte or eight bits) is sent with no separation between characters. In order for the receiver to be able to define the boundaries, the transmitter must send one bit with every clock pulse. Then every eighth pulse constitutes the last bit of the data unit. (Please note that data units can be five, six, seven, eight, or nine bits, so the sending

Figure 6-2. Standard RS 232 Pin Assignment

Pin No.	Use	Data Direction
1	Chassis Ground	Common
2	Receive Data	In
3	Transmit Data	Out
4	Request to Send	Out
5	Clear to Send	In
6	Data Set Ready	In
7	Signal Ground	Common
8	Data Carrier Detect	In
20	Data Terminal Ready	Out
22	Ring Indicator	In

and receiving computers must both have the data unit length set alike.) The stream is preceded by either one or two synchronizing characters which prepare the receiving computer to receive the data. The data follow, one data unit after the other, until the end of the message. This is a simplified explanation of a complex series of electronic signals. The following is an example of what occurs between two computers using synchronous communication:

The transmitter sends an enquiry character to determine if the receiver is ready to receive.

If the receiver is ready, it sends back a character that says ready.

The transmitter now sends a character that tells the receiver that what follows is data. It sends a start of block character, then the actual block of data, and finally an end of block character. The receiver checks the block using various error-checking techniques to determine if transmission errors have occurred.

If errors are not detected, the receiver sends a character that signals it is all right to proceed with the next block.

The transmitter sends the start of block character and the next block followed by the end of block character.

The receiver checks the data as before.

This cycle repeats until all blocks transfer and then the transmitter sends the end of data character.

Within synchronous data transfer, there are a number of data organizations or protocols used. The first and most common is the Bisync protocol. Bisync identifies the beginning, the intermediate, and the end of messages using special control characters. This is the protocol used by most current CAD/ CAM system computers, although new protocols are rapidly making inroads. SDLC, the Serial Data Link Control, and HDLC, the High-Level Data Link Control, protocols permit transmitting data as a bit stream of any length up to the maximum a memory unit can hold. They also dispense with implied character boundaries. However, details are unnecessary here because the essential point is that several protocols exist to link computer systems. Computer acquisition projects must include an activity for working on integrating data between new and existing equipment. Detailed discussions of synchronous protocols are found in the following publications:

1. IBM Synchronous Data Link Control General Information GA27-3093-1
2. IBM 3650 Retail Store System Loop Interface OEM Information GA27-3098-0

Digital Equipment Corporation has developed the Digital Data Communications Message Protocol (DDCMP). Some CAD/CAM systems are starting to use this protocol to interface CAD/CAM systems to Digital Equipment Corporation computers. Synchronous data transfer is the preferred method for computer-to-computer communications for two reasons—it's fast and it's efficient. It is the inherent compactness of data using this method that makes it fast. This speed, however, makes synchronous communication difficult if data transmission lines have any significant electrical noise on them. Significant costs are incurred to obtain noise-free transmission lines.

The second method of transfer is asynchronous. This method differs in that a character transfers when the transmitter has a character to send. In between characters, the transmitter sends a continuous mark signal. This means that each character sent must have a start bit to identify that a character is coming and either one, one and one-half, or two stop bits follow the character to tell the receiving computer "you got it all." Thus, each character is framed rather than simply framing a whole data block as in synchronous. The start and stop bits required to send messages slow transmissions and produce inefficiency. This is the method normally employed to do microcomputer communication over telephone lines to time-sharing computers. The timing needed by the synchronous method is critical and it is easier to use this asynchronous method when timing can be a problem. Telephone voice grade phone lines are too noisy to permit synchronous data transfer. The capital costs associated with asynchronous communication are lower than those of synchronous. Low data volumes transfer more cost effectively using asynchronous means. Because of the lower speed and the individual character framing, noise levels on transmission lines can be higher and still transfer data error free. Special noise-free transmission lines are unnecessary, so line costs drop.

CHARACTER SETS

Computers send bits that convert to characters and numbers, which are meaningful to human beings. Computers like humans have different character sets and to forget that is tragic. Chinese, Arabic, Russian, and English all have different characters. Computers normally have ASCII, the American Standard Code for Information Interchange, but many have EBCDIC, the Extended Binary-Coded Decimal Interchange Code. ASCII is the standard for almost all 8-bit microcomputers and for many mainframes as well. EBCDIC was developed by IBM Corporation and is the standard on most, if not all, IBM mainframe computers. We have another potential problem; computers use different character sets as well as different programming languages. Computer character sets are arbitrary conventions as are human language character sets. There are routines incorporated in computer programs that convert EBCDIC to ASCII and vice versa. These converters also permit communication between computers when the character sets differ.

NETWORKING SYSTEMS

Networking is the term applied to the linking of computers to permit data transfer and sharing. The physical interconnection of computers can be in one of three possible configurations: star, ring, or chain. The star configuration is essentially a central primary computer at the core of the star connected to satellite computers. Figure 6-3 shows this configuration.

The star network has the primary advantage that it is faster under normal conditions than the ring or chain. It has two major disadvantages: the primary

Figure 6-3. Star Network Diagram

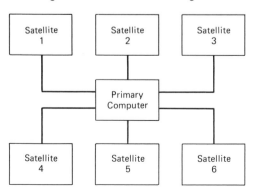

slows down when demands for data are maximum, and the entire network is dependent on the primary system. In the event of primary system failure, the entire network goes down until the primary is fixed. The star network has the primary data base connected to the primary computer, so interruption of the primary causes complete system shutdown. This can be a significant problem for businesses that are entirely dependent on their computer systems.

The chain network, consisting of a series of computers connected in a line, assumes all computers are equal in status and data are distributed. This configuration prevents failures from shutting down the entire system, but it takes longer to get data from one end of the chain to the other. This configuration is diagrammed in Figure 6-4.

There is no theoretical limit to the number of computers that can be connected serially, but the software is more complex and the data delays long when the chain is long. Failures in the chain isolate sections of the chain.

The ring configuration corrects the isolation problem and improves the response of a chain. The ring is a chain that has had its ends connected. Figure 6-5 shows a diagram of the ring configuration.

If any single computer fails, the others can all communicate as they would in a chain with only the data in the failed computer unavailable to the others. The data path in this configuration is half as long as in the chain between remote computers. Thus the time for data transfer is halved for rings with the same number of stations as the chain.

The performance of each of these networking configurations is dependent upon the choice of communications protocol and network philosophy. It is best to study the response times, throughput data loads, and effects on company operations of failures to choose the best configuration for a given application. This analysis is best done by data processing professionals, but the final decisions are most appropriately made by informed, educated users. Learning about alternatives has two benefits: (1) Users get a voice in the system operation that they will be using, and (2) in event of failure in the network they know what to expect and how to compensate. This is vital, as we all know from practical experience with Murphy's law! The ease with which users compensate for interruptions is crucial to business efficiency.

The major benefit to networking is that a single computer failure doesn't shut down all operations. The other computers keep functioning, and only the

Figure 6-4. Chain Network Diagram

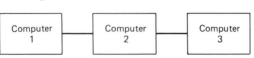

Figure 6-5. Ring Network Diagram

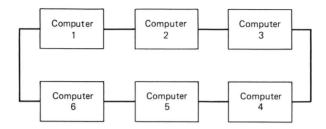

operations on the inoperative processor need be expedited when repairs are complete on the system. If network capacity is adequate, operations from down computers shift to another host computer in the network and no loss of time occurs.

INTEGRATION TO BUSINESS MAINFRAMES

CAD/CAM systems and engineering computers can be integrated with business mainframes in several different ways. The easiest way is to permit the engineering computer to play the role of a terminal to the mainframe. This permits access to the mainframe files using the conventional programs virtually without the need of complex file transfer software. This gives engineering access to the data without, necessarily, the ability to change data on file and with a minimum investment in computer software. There are several drawbacks to this system, however. Engineering would have to get a hard copy of data or manually transcribe data from the system to use within their own computer. The engineering group could also obtain the necessary software to capture data sent and simply put it in a disk file in the form received. This can be done using asynchronous communication software without involving complex networking systems. This method is practical for some kinds of data and is the minimum communication tolerable, but it is not an adequate integration of systems.

The second level of integration—only one step above using the engineering computer as a terminal—is to provide both computers with either synchronous or asynchronous data transfer in both directions. This capability allows engineering to capture segments of the main system files for analysis or change and gives engineers the ability to send the changes or updated information back to the main system in batch mode. This ability to batch data back and forth is the minimum communication necessary to call the system integrated, because it permits engineering and data processing to take full advantage of electronic data transfer in two directions with the ability to operate both systems totally independently. It also allows file sharing without the need to address the mainframe data base each time to get data. This method is not without problems, however.

The data in the main data base that engineering is to capture must be software protected in certain fields. For instance, the payroll records needed to get labor cost data are confidential. Trouble arises when engineering needs to access certain data that are stored together with payroll and personnel records. Adequate data security must be maintained. On the other hand, engineering needs to be able to make changes to material specifications, bills of material, routings, tooling data, and the like and to update the main records. So software must prevent tampering with but still give access to some data elements. This software is part of the cost of integration and is included as part of the integration project.

The third level of integration is to network the computers together to allow engineering to keep drawing files but to maintain bills of material and routings on the main computer. This concept of full integration minimizes the amount of file duplication, but it necessitates the use of considerable software in both the mainframe and the engineering computer. Full integration provides for the needs of both with their separate languages and permits minimum file storage for both computers. It is the best way to maintain accurate records with current data available to all users. However, it has the same security problems as the batch update, and software for this level of integration is complex and expensive.

WORK STATION LINKS

Almost all 16-bit minicomputer systems and even many 8-bit microcomputers permit multiple users. This allows the sharing of data, printers, and processor, which reduces the cost per user of the system's resources. As with mainframes, the more users a system has, the greater the utilization of those resources until a saturation point is reached. Saturation doesn't result in denied access to the computer. The system will do as manufacturing does when it reaches maximum capacity; it delays execution of all jobs, establishes priorities, and reduces productivity. Thus, there is a finite limit beyond which having additional work stations is undesirable. That limit is the system saturation point.

The saturation point is defined in terms of response time—the time it takes to get a response to an input. The minimum tolerable response time is subjective and application dependent. Therefore, the number of work stations supported by a given computer or CAD/CAM system before saturation can vary widely. There is a rule of thumb for graphics and computations that works well in practice: Try not to exceed one quarter the number of bits in the processor for the number of work stations involved. For example, using more than two work stations per 8-bit processor or more than four stations on a 16-bit processor produces significant speed reductions. This does not mean that it isn't possible to add more than that number of work stations, but it does mean that difficulties are avoided by remaining below that number. As with most rules of thumb, it is only a guide. In general, don't plan to connect more than five terminals to a 16-bit processor used for drafting or mathematical analysis if they are to be operated simultaneously or extraordinary delays will result. If the terminals operate at different locations and not simultaneously, then there is no reason to limit the number except to establish rules for the number in use during any period. There are exceptions to "the one quarter rule." Some processors, for example, use other processor chips to control peripheral devices and disk drives and to handle input/output. These computers can support a greater number of terminals without speed loss.

WHAT DOES ONE DO WHEN UNINTEGRATED HARDWARE ALREADY EXISTS?

It isn't necessary to buy new computers to get existing computers to talk to one another. Computer suppliers are sensitive to the need to integrate systems and can advise their clients on what it takes to integrate systems already in existence. An important first step for a company with a patchwork of computers is to educate someone who knows the company and the existing systems in the integration alternatives. This ensures that business, data security, and throughput needs are all considered. Outside consultants can do the analysis, but company policy is usually best set by in-house personnel who know the needs and can provide the judgments necessary to maximize return and minimize disturbance.

INTEGRATION PROJECT

Perhaps the best method of achieving an integrated system is to establish an integration project team. The team will be charged with the responsibility of integrating systems from desktop personal computers to CAD/CAM and business mainframe computers. It should consist of data processing professionals along with representatives from engineering, accounting, personnel, materials control, manufacturing, and top management. The team's task is to produce an overall integration plan to balance the resources and the needs of all groups involved with the technical possibilities needed to achieve integration. In addition, the team should also delineate the policy for acquisition of future additional computer hardware such that integration is accomplished with minimum strain. These guidelines for the future should provide maximum flexibility in hardware and software choice for the user and not preclude integration or equipment acquisition.

Once policy is set, the team should decide upon an integration project manager to execute the plan. Committees execute poorly because routine decisions take too long. Project managers execute plans best. The project manager makes all routine decisions necessary to implement the plan. The committee oversees the project to eliminate violations of company policies and advises the manager on nonroutine decisions. The project manager should have full authority and be the focal point of communication on all facets of the project. Project planning and control is covered in Chapter 9.

INTERFACING WITH ANALOG DEVICES
AND PROCESS CONTROLLERS

Internally, computers use digital or discrete data bits. Most real world phenomena, such as temperature, pressure, humidity, displacement, and velocity, are analog-measured information. A conventional glass tube thermometer has an analog display. Many automobile speedometers and clocks have analog displays. Control devices use infinitely variable voltage or current to adjust processes. The interface changes these analog signals to digital data. This is an

Figure 6-6. Typical Measurement and Control Loop Diagram

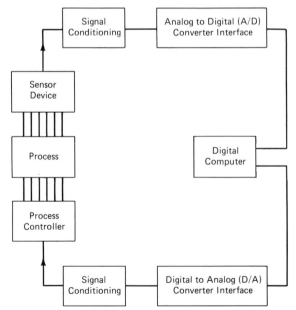

A/D or analog-to-digital interface. Figure 6-6 is a diagram of a typical measurement and control loop.

Signal conditioning translates converter interface signals from sensors or to controllers. A/D and D/A converters need input and produce output in ranges far from that required by controllers or produced by sensors. The input signal conditioner filters the input, provides power surge protection, and translates the signal from the sensor to the signal range of the A/D converter. The output signal conditioner works on the output of the A/D converter to make it compatible with the process controller.

COMPUTERIZED PROCESS MEASUREMENT

Computers are now reading gauges and processing the information they receive. The key to successful computerized process measurement is careful planning. Planning determines

1. How frequently the data signal is sampled
2. What resolution is required and available
3. What computations will be done by the A/D converter
4. What computations will be performed by the computer

Computerized measurement systems sample analog sensors for instantaneous data. To make an exact record of a varying signal requires noting the value at every instant. This is not practical, so sampling the signal frequently provides a statistical approximation of the signal. The Nyquist sampling theorem recommends a sampling rate of at least twice the signal waveform frequency. For a human blood pressure waveform of 3 Hertz and a fifth harmonic resolution, the estimated minimum sampling rate is 36 samples per second.

Setting the sampling rate is critical for signal approximation, but higher sampling rates are not necessarily better. Doubling the sampling rate doubles the volume of analysis data. It also doubles the overhead of computer time to save data from the A/D converter. This reduces the number of sensors simultaneously monitored by a given computer.

Resolution is a consideration for the sensor, the A/D converter, and the sampling frequency. Sensors with outputs between 0 and 0.75 volts combined with an A/D converter with a resolution of 0.5 volts won't provide much useful data on a varying signal. A 12-bit A/D converter has a resolution of one part in 2048, or 2.12. It is inappropriate for a required signal resolution of 1/4096. In this example, if the 12-bit A/D converter had a ± 10 volt output, the resolution would be 4.9 millivolts. On a 0 to 0.5 volt signal, it provides 1 percent accuracy. It is very difficult and expensive to obtain accuracies above

99.9 percent. Noise, drift, and nonliner inconsistent sensors and circuits are a constant problem.

Transient analysis is a real benefit of digital processing. Digital data are accurate and stable, and stored data can be manipulated with no loss of accuracy. In many situations, the desired signal is obscured by noise, and filtering won't eliminate the noise. The digital average transients technique employs multiple readings of the same signal to remove the background noise. The signal-to-noise ratio improves in proportion to the square root of the number of observations.

REAL-TIME VERSUS BATCH PROCESSING

The point of calculation is determined after setting the sampling frequency and the accuracy requirements. Which calculations need to be done in real time and which can be stored for later processing? Feedback results used at measurement time must be in real time. These may be feedback signals or display data. The following points are pertinent for less clear situations:

1. Batch processing permits unanalyzed data to be stored.
2. Real-time processing reduces some data to results that preclude further processing at a later time.
3. Batch processing permits analysis, which takes too long in real time.
4. Real-time processing reduces the quantity of data and the space on mass-storage devices to retain it.
5. Real-time processing reduces the processing needed later.

A good general candidate for real-time preprocessing is any experiment with over 100,000 data points per hour. An assembly language programmer is useful for real-time processing. Assembly language is the language of real-time programs. Batch type programs, written in a high-level language, run slower. Real-time programs must be fast enough to process the data in real time. Assembly language programs are frequently the only ones fast enough to do the work in the time between samples.

7

PART NUMBERING AND CODING

No discussion of engineering's role in manufacturing is complete without part numbering and coding. Every company has some form of part numbering or coding system that works more or less well, and engineers generally give these systems little thought. However, the structure of numbering systems and the way they work has changed dramatically in recent years. Thus engineers need to reevaluate existing systems periodically.

The importance of coding systems to enhance organization and the orderly operation of business has been recognized since the early 1900s. H. K. Hathaway, who had worked with Fred Taylor, wrote the following in 1920:

> A classification is essential to an orderly arrangement of the facts relating to a business and to the orderly conduct of its activities. During the period of development and installation of a system of scientific management, it is especially helpful to a proper visualization and understanding of the business and its problems as well as to the conduct of the work.

This statement is as true today as it was in 1920. Scientific materials management requires part numbers or codes to identify each item used. Accordingly, the first requirement for designing a part numbering or coding system is to define the purpose it is to serve. The primary objectives are:

1. To identify the item
2. To permit easy storage and retrieval
3. To shorten time in record keeping
4. To establish a classification system

The code or part number must establish a unique identifier for the item to prevent part confusion. The unique identifier permits easy storage and retrieval of data on the item in either a manual or computerized system. The part number is usually brief in comparison to the name or description of the part. The identifier need not describe the part using a feature code.

The first three objectives are essential to any part coding system, whether in a computerized or manual system. The fourth objective provided order and error checking in the pre-computer age. When inventory records were kept manually and fewer active parts were made, it was useful to have part codes with descriptive information. Even in the early days of punched cards, sorting cards on two fields was impractical. Putting the descriptive data in the part number incorporated parts classification. Then sorting the part number file manually completed part number file maintenance. The file contained all data used by both inventory control and the stockroom. Since the classification most frequently desired was based upon end product shape or physical feature information, the part could be identified by using the part number alone. This provided error checking.

A computer system sorts a data file on any field or sequence of characters easily, and the classifications may be put in a separate field from the part identification code or part number itself. This means that parts may be identified or part numbered using any assigned series of characters or numbers without regard to order or sequence, and easy sorted classifications can still be maintained. Descriptive groupings are obtained by sorting a classification field. Today's computer systems make possible the separation of part numbers and classifications. This independent classification system made little sense in the pre-computer age because dual coding of parts was difficult additional work. The major advantage to separate fields for identification and classification is that part identification numbers can be shorter and classifications, more complete and extensive. Many companies still maintain elaborate systems to classify and identify parts. These companies carry long part numbers into computers, but they pay a price to retain that system.

PART IDENTIFICATION SYSTEMS

Part identification systems fall into six groups:

1. All numeric, nonsignificant
2. All numeric, significant
3. All alphabetic, nonsignificant
4. All alphabetic, significant and usually mnemonic
5. Combinations of letters and numbers, nonsignificant
6. Combinations of letters and numbers, significant

The all-numeric system identifies the part with a unique numeric identifying number without regard to classification of the item. General Motors uses a six- to eight-position numeric part numbering system for part identification.

The most common all-numeric significant part numbering system is the Dewey Decimal system employed by libraries for book coding. Even this system uses a letter or alphanumeric character to separate authors. The fundamental Dewey system for book classification is numeric with alphabetic added to separate books having the same classification but different authors. Since the Dewey system developed when fewer books were available, the alphabetic addition is really a work-around designed to permit the original Dewey system to be used in an expanding book universe. The Library of Congress uses its own classification system developed because the Dewey system was inadequate for very large libraries.

The all-alphabetic systems are similar to the all-numeric systems but the major advantage to letters is the 26 possible different characters per position that this system affords. For a six-digit numeric code, the number of possible different parts is one million. For a six-character alphabetic code, the number of parts is over 300 million. Although this system has been tried in manufacturing companies, it has been abandoned in favor of a combination numeric and character system.

A significant letter or mnemonic item numbering system was proposed by Gilbreth for tool coding. This system is best described by example. Suppose a standard 0- to 1-inch micrometer caliper needs a descriptive code. Under Gilbreth's system the code would be:

$$M \;=\; \text{Measuring Device}$$
$$C \;=\; \text{Calipers}$$
$$E \;=\; \text{External}$$
$$C \;=\; \text{C Type}$$
$$A \;=\; \text{Adjustable}$$

Thus, the micrometer has the number MCECA with a 0- to 1-inch range usually specified as a numeric suffix (MCECA 0-1).

The combinations of letters and numbers or the alphanumeric nonsignificant system is best demonstrated by automobile license tags. Most of the states have no significance to either the characters or the numbers used. There is frequently some pattern significance to distinguish between automobiles and trucks in the form of the number of letters and their position but within the passenger car population most tags are nonsignificant. An example of pattern significance is as follows:

Passenger cars have the pattern XXX-000
Trucks or commercial vehicles use XX-0000

where X represents any letter and 0 any number.

Even this system is not pure because certain three-letter combinations are omitted to prevent them being interpreted as words. States also issue letter groups to counties so the letters indicate the county of residence. The letter omission example was demonstrated in Michigan in the 1970s when polybromated biphenyls contaminated some farms. Some people turned in their plates to get rid of PBB as the first three letters. States omit all the dirty three and four letter words and those that phonetically duplicate these well. Profanity is also dropped.

The system in widest use in industry is the combination letter and number significant part numbering system or some slight variation on significance. The bearing industry uses significant part numbering. The letters indicate the type and the numbers reflect the bore and outside diameter of the bearing. American Motors uses one or two digits as significant with the rest nonsignificant. Ford uses a 22-digit significant alphanumeric system.

The Ford system is composed of a 7-digit prefix, an 8-digit base, and a 7-digit suffix which is broken down as follows:

Seven-digit prefix—Example: ED2E4AB

First three digits	Engineering Deviation Code	(ED2)
Next two digits	Year (1984)	(E4)
Next digit	Carline (Ford)	(A)
Next digit	Design Responsibility	(B)

Eight-digit base—Example: 54200A10

| First two digits | Body Type (four door sedan) | (54) |
| Next six digits | Basic Number | (200A10) |

Seven digit suffix—Example: A0000AZ

First digit	Design Level	(A)
Next four digits	Alternate Parts Code	(0000)
Last two digits	Change Level	(AZ)

Total part number ED2E4AB 54200A10 A0000AZ

Thus, there are as many types of systems as there are ways to number and code parts. The point is to determine what purpose the system is to serve and note the advantages and disadvantages of each alternative as it relates to that purpose and use within the manufacturing plant.

The number of digits in the part number is irrelevant for engineering purposes or for computer system purposes, but length has an effect on human reporting accuracy in the manufacturing plant. The longer the part number, the greater the probability of errors when people use it. Studies by psychologists have demonstrated that the reporting error probability increases 5 percent for each digit added beyond six digits. Six-digit numbers and smaller have about the same reporting error frequency. Thus, a 12-digit part number will have a 30 percent greater error probability than a six-digit one in factory use.

High error rates reporting long part numbers has been a recognized problem for some time. Systems have been designed to reduce the problem. Prepunched IBM cards, bar codes, optical character readers (OCR), and other high tech systems are being used to minimize errors. Computers provide check digits and are programmed to reject some of the errors as well. The real question is not how to get accurate reporting of long part numbers but how to make part numbers shorter to reduce the errors and still satisfy the original purpose for significance.

PART NUMBERING ELEMENTS

A part numbering system should be designed to take into consideration the following five vital elements:

1. Minimum number of characters: Part numbers should be as short as possible.
2. Expandability: A company will grow and need more part numbers. The system must support additional numbers as required.
3. Flexibility for new options: The system must support additional options as easily as it supports additional numbers.
4. Ability to be selected and sorted: Although classifications may or may not be included in the part numbering system, the records on the parts must sort easily into classifications.
5. Ease of assignment: Numbers should be easy to assign to new parts.

In large organizations, estimates of part number error costs for single errors exceed $10,000. Strict adherence to an engineered system pays. This high cost is a clear incentive to reduce error rates. The reason for these high costs are:

Parts ordered to replenish stock not really used causes:

1. Excess inventory
2. Increased inventory carrying cost
3. Larger warehouse needs

Parts not ordered and available for production causes:

1. Idle workers and machines
2. Overtime when the parts arrive

3. Airfreight costs
4. Late shipments
5. Reduced customer service
6. Substitution of higher cost items
7. Higher cost alternate operations
8. Higher freight costs

PART CLASSIFICATION

These requirements in the pre-computer age produced two fundamental classification systems:

1. Classification by likeness
2. Classification by function

In classification by likeness, all items are grouped by variety into some general classification. For example, all drills are grouped together as a class although some were brad point and designed for wood and some were twist, high-speed for metal.

The function system was developed to classify parts by function or use rather than by likeness. Frederick W. Taylor described this system and established general tool groups. The function class in which drills fit was classified under D—Drilling and Boring Tools. This functional group contains drills, reamers, boring bars, cutters, taps, and broaches.

SIGNIFICANT PART NUMBERS

In a small company with only a few parts to classify, a classification system can be both simple and easily incorporated into the part number. This reduces manual record keeping and is a productivity improvement program. The significant part number can also be used to check the physical part against the number to eliminate reporting or manufacturing errors. For example, a Torrington B1620 needle bearing has a drawn cup, one-inch bore with a 1-1/4 inch length. The B indicates the drawn cup type. The first two digits represent the bore size in 16ths of an inch and the last two digits define the length in 16ths.

As a company grows and the number of parts increase, the classification system becomes more complex. Then the significant part numbering system becomes more useful from the point of view that the part number actually describes the part. People familiar with the system recognize when the part they have doesn't agree with the number. For example, a Torrington bearing, with the part number B1624, will be a drawn cup, full compliment, trunnion end needle roller bearing. This is defined by the B. It will have a one-inch bore and a 1-1/2 inch length. It is easy to tell that you have an incorrect part if the bearings in the shop box awaiting grease have a half inch bore and they are labeled as B1624's.

Significant part numbering systems are not without problems, and the first of these is the limitation in the maximum number of parts classed through the part number. Since each field or range of digits in the number must contain all possible combinations of that particular characteristic, the design of the system must anticipate all the combinations possible or build the system for longer numbers. If a company limited itself to a five-digit bearing part number with the first digit an alphanumeric character, it could only make A through Z types with bores and lengths limited to 99/16ths inches before significance disappears. But the fact is that companies do make new parts even if those parts scrap the significant numbering system.

A second problem encountered is that the number must grow to handle options; the reporting error rate increases with the length of the number. Although significance attempted to improve productivity and reduce errors, it deteriorates over time to reduce productivity and increase errors. Error rates are higher for those that must communicate numbers and alphanumeric characters by telephone. The phonetic alphabet shown in Figure 7-1 is used by the military to help reduce orally transmitted alphabetic characters. Some numbering systems prohibit the use of the letters O and I because of the written similarity to zero (0) and one (1). An upper and lower case letter system drops the lower case L because it has been used by typists as a 1 for many years.

Figure 7-1. Phonetic Alphabet

A – Alpha	N – November
B – Bravo	O – Oscar
C – Charlie	P – Papa
D – Delta	Q – Quebec
E – Echo	R – Romeo
F – Foxtrot	S – Sierra
G – Golf	T – Tango
H – Hotel	U – Uniform
I – India	V – Victor
J – Juliett	W – Whiskey
K – Kilo	X – X-ray
L – Lima	Y – Yankee
M – Mike	Z – Zulu

The wide use of computers has produced another problem. To keep records in a computer, the number of characters in a field is predefined as part of the software installation. A company that elects to use a significant part number system has a major software and file revision task if it needs to expand predefined length at a future date. Some of the new data management systems have made changing field lengths easier, but significant program revisions are usually necessary as well.

NONSIGNIFICANT PART NUMBERS

Nonsignificant part numbers are sequence numbers. They are assigned in sequence to parts added to the system. Because they are assigned in sequence, they have no classification built into them. This is the first advantage of nonsignificant systems—flexibility. Because nonsignificant systems function independently of characteristics or standards, any part can be updated, changed, or revised as the need arises.

The second major advantage to a nonsignificant system is that a given number of digits in the part number can handle more parts with greater flexibility in options than any significant numbering system. Take the simple five-digit Torrington bearing system on p. 95 for example. The A to Z first character permits 24 different types if we eliminate O and I. For nonsignificance the first digit could be a character and have 24 options as well. The nonsignificant system would assign all As before using Bs.

The next two digits can have a maximum of 98 possible numbers because 00 is without meaning. The same holds for the last two digits. The significant system can cover 24 times 98 times 98 parts or 230,496 possible parts. The nonsignificant alphanumeric system can have 24 times 10,000 or 240,000 parts. Options are not limited to even 1/16 inch increments and more parts in total can be numbered.

The third advantage of nonsignificance is that it covers all combinations of hardware and parts regardless of class of product. It also eliminates duplication and having a dual system. Few significant numbering systems accommodate finished product as well as the purchased components used to make it up. Thus, purchased parts usually have a second numbering system quite different from the finished parts system used. Nonsignificance permits a single system for all parts.

The fourth advantage is really the lack of a specific disadvantage. Significant part numbering systems in companies with large numbers of items are simply too complicated for most workers to memorize well enough to check errors as these systems were originally designed to permit. This means that error checking cited as an advantage is rarely effective. The nonsignificant system never had an error handling capability and it has fewer errors by being shorter to report.

Nonsignificant systems are not without problems, however. The fundamental problem with nonsignificance is the inability to segregate major product lines or classes of products by simply using the part number. This can be accommodated by using a partially significant and partially nonsignificant numbering system. Classification can also be accomplished by putting group code data in another data field in the computer record. This provides the benefits of some significance to separate major product groupings in the part number with the benefits of a nonsignificant numbering system. This is a semi-significant part numbering system.

THE SEMI-SIGNIFICANT NUMBERING SYSTEM

The most common semi-significant numbering system is the aluminum alloy group designations. Groups use the first digit to identify the major alloying element with the second digit being used to indicate alloy modifications, but the next two digits are nonsignificant. The standard aluminum designations are:

1XXX	99.00% and higher purity
2XXX	Copper alloy
3XXX	Manganese alloy
4XXX	Silicon alloy
5XXX	Magnesium alloy
6XXX	Magnesium and silicon alloy
7XXX	Zinc alloy
8XXX	Other element alloy
9XXX	Unused series

Thus, a 6061 alloy is a magnesium-silicon aluminum alloy. Since the second digit is a 0, it is the first or original alloy produced in this group. A 6101 alloy has the same basic alloying elements, magnesium and silicon, but is the first modification of the original alloy (61XX) as designated by the last digit being a 1. So, there are benefits of semi-significant numbers in the aluminum industry.

The semi-significant system has most of the benefits of a nonsignificant system and provides family identification as well. This is the system toward which most large corporations move when faced with the problem of having to revise a significant system because they have run out of available numbers. Although the most flexible is the nonsignificant, the semi-significant offers a potentially valuable compromise. Avoid proliferation of prefixes and suffixes or this proliferation leads to semi-significant systems deteriorating into significant ones.

GROUP TECHNOLOGY CODES

The biggest problem with the nonsignificant numbering systems is the loss of characteristic data in the part number. In the days of the index card filing system, this was a problem because more data had to be maintained on the card so the file would sort by classification. Today sorting is no problem. Sorting in a computer on a field other than part number is easy. This fact creates controversy today because manufacturing knows a group technology field in the part record is possible. Engineering frequently desires significant part numbers and they control part numbers. The controversy arises because data processing and engineering have given little consideration to the needs for a group technology field in the record. Manufacturing desires short part numbers for accurate reporting, and separating groups from the part numbers has not been thoroughly researched.

The group technology code contains all the material, dimension, type, process sequence, and tool requirements for manufacturing the part. This GT code may be similar or identical for a number of parts without causing problems because it isn't the part number and does not need to be unique for each part. This field, sorted by the computer, produces lists of parts using the same processes, materials, dimensions, tools, or work centers for purposes of developing standardization of parts and the manufacturing process, and standardization is the key to minimizing inventory and maximizing productivity. The better and smoother the work flows through the work centers in the plant, the lower the inventory and the higher the output of product per worker hour in the plant.

COMMON GROUP TECHNOLOGY CODING SYSTEMS

Monocodes

An understanding of the inner workings and structures of GT coding systems will help put them to use effectively. The oldest type of group code is the biological classification system of Linnaeus. This structure is:

Kingdom
Phylum
Class
Order
Family
Genus
Species

This type of structure is a tree or hierarchical structure. In GT, this type of code is a monocode because each digit or character has a singular (mono) meaning. Each digit or character has significance only if all characters to its left are known. Coding a part, given the structure, is easy: Start at the top or trunk and work down level by level to the end. Carried to the extreme, this system produces singular identification of items as in the biological classification system. No two plants or animals have exactly the same classification. This system produces a significant numbering system if taken to its limit.

Polycodes

The second major type of GT classification system is the polycode system. In this system, each digit is significant by position alone. Each fundamental group or feature is defined by a single position in the number. The difficulty with polycodes is that they require long numbers to describe all part features.

Every polycode system developed to date is positional dependent because digits appear in groups. Diameters are specified using more than one digit as are most other dimensions. This produces a group significance independence with digit interdependence within the group.

Hybrids

Most industrial GT coding systems are hybrids of the monocode and polycode systems. This gets around the length problem of the strict polycode but retains the polycode features. This type of code structure is a semi-polycode. It divides the total population into a number of subgroups with the leftmost digit or digit series and proceeds from there as a polycode.

Group Technology Codes

Although there are many custom codes developed for specific applications, there are a few common commercial codes in use in many companies. Many custom applications are products developed by modification of these basic commercial systems.

To compare these systems, a common washer-shaped part will be used. The part is shown in Figure 7-2.

PNC (Brish-Birn type developed for General Motors)

This is a six-digit monocode system.

Code is 311312

3—Piece part

Figure 7-2. Part Drawing - Special Washer

Simple Example of a Coded Part

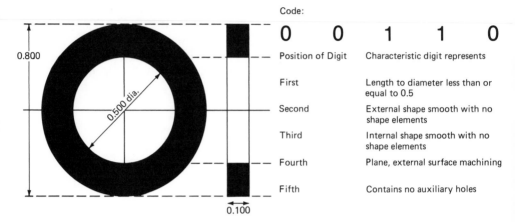

Code:				
0	0	1	1	0

Position of Digit	Characteristic digit represents
First	Length to diameter less than or equal to 0.5
Second	External shape smooth with no shape elements
Third	Internal shape smooth with no shape elements
Fourth	Plane, external surface machining
Fifth	Contains no auxiliary holes

Note:
Based on the first five digits of the
Opitz coding system

(Courtesy of George Plossl Educational Services, Inc.)

1—Round with straight centerline
1—Metallic, short, plain (no threads, splits, etc.)
3—Single through centerhole without other holes
1—ID greater than 0.500 and less than 0.570
2—OD greater than 0.800 and less than 0.900

OPITZ (Dr. H. Opitz, Aachen, West Germany)

This is a nine-digit semi-polycode.

Code is 001101610
0—Round part with L/D ratio < 0.50 with straight centerline
0—Single OD without threads
1—Single ID or stepped to one end without threads
1—Planar machining of faces, but no slots or grooves
0—No auxiliary holes or gear teeth
1—OD greater than 0.800 and less than 2.00
6—Material type and heat-treat condition

1—Initial material form
0—Tolerance class

CODE (Manufacturing Data Systems, Inc., Ann Arbor, MI)

This is an eight-digit hexadecimal semi-polycode.

Code is 11200061
1—Round part concentric about straight centerline having no teeth, splines, or gear teeth
1—Single OD
2—Single diameter through centerline without threads
0—⎫
0—⎬ No flats, slots, protrusions, grooves, or holes
0—⎭ except for centerhole
6—OD greater than 0.72, less than 1.20
1—Length greater than 0 and less than 1.00

MICLASS (Dutch Metal Institute distributed by TNC, Waltham, MA)

This is a 12-digit semi-polycode.

Code is 112022112133

1120—Main shape and shape elements
2211—Dimensions
  21—Tolerances
  33—Material

TEKLA (Norwegian NAKK Project)

This is a 12-digit hexadecimal semi-polycode.

Code is 00X220340000
00X—Cylindrical shape made from round bar stock
220—Plain single OD and ID
340—Two opposite axial faces machined and without gear teeth
000—No auxiliary holes

A comparison of these system is shown in Figure 7-3. Figure 7-3 also lists the information contained in each of these coding systems.

Figure 7-3. Coding System Information Comparison

	P N C	O P T I Z	C O D E	M I C L A S S	T E K L A
End Shape	X	X		X	X
Outside Shape	X	X	X	X	X
Inside Shape	X	X	X	X	X
Protrusions		X	X	X	
Additional Holes	X	X	X	X	X
Threads	X	X	X		X
Grooves or Slots	X	X	X	X	X
Flats	X		X	X	X
Gear Teeth or Splines	X	X	X	X	X
Splits, Keyways, Knurls, or Swages	X				
OD range	X	X	X	X	
ID Range	X				
Length Range	X	X	X	X	
Size Ratios	X	X	X	X	
Tolerances		X		X	
Heat Treat		X		X	
Material Form		X			
Material Type		X		X	
Finish				X	

These group technology codes provide part form information and contain very little about how the part will be manufactured.

COMPUTER-AIDED PROCESS PLANNING

If a part has been classified using a GT part coding system and process routing data have been coded for each part, a complete feature/process history is available. These data provide a basis for searching the routing files for common routings for parts with similar features. The goal of searching the data is to determine common routings for similar parts given the GT feature code alone. Common routings simplify routing development, improve production times, and reduce work-in-process inventory.

The MICLASS coding system is a 30-digit GT coding system which provides complete part features in the first 12 digits and offers the remaining 18 for company-related information. These include common lot size, piece time, major machining operations, vendor codes, and in-plant process data. There are other standard systems available today but the ultimate goal of any system is to improve the flow of manufacturing work. The data in any GT system must be available for manufacturing as well as engineering. The typical savings are as follows:

Manufacturing process planning	58%
Direct labor productivity	10%
Quality improvement	10%
Improved tooling	12%
WIP inventory reduction	10%

Manufacturing process planning or routing development consumes considerable engineering time. This time is dramatically reduced by having the computer suggest a common routing for consideration and revision. In some cases, the computer-generated routing is fine. At other times, the computer routing must be modified. In all cases, it is easier and faster to start with a suggested routing rather than having to start with a blank page.

Direct labor productivity rises because of common routings. The ability to make common part families using a standard routing gives production workers practice doing jobs over a consistent process path. This assists workers in learning the standard task elements and, over time, results in applying time formerly used to learn a setup to actual production.

Quality improvements occur because standard routings result in standard tooling. Standard tools can be stocked or produced as needed more easily. The combination of standard tooling and improved worker training on standard routings results in fewer scrap parts and improved process quality. Standard tooling is cheaper and easier to make and to service. This provides lower cost tools at the same time as it increases quality.

With higher product quality, better material flow, and improved productivity, the work-in-process inventory drops. Although this WIP reduction is not automatic nor guaranteed, management can see a more reliable workplace which needs a lower WIP level. As problems are solved, the inventory used to insulate the plant from them can be eliminated. This is part of the Just-In-Time philosophy covered in Chapter 15, "The Just-In-Time Philosophy."

BAR CODING

The great effort today is to minimize errors in factory data reporting. This same effort has been under way in the grocery industry for some time. Bar coding started in 1932 with a Master's Degree thesis at the Harvard Business

School that suggested grocery store checkout be automated. This idea rested for over 20 years until in the mid-1950s Philco and Sylvania conceived the idea of a scanner to read a code imprinted on the products sent through grocery checkout registers. Although a recognized good idea, it again rested until 1967 when a pilot scanner system was installed in a Kroger store in Cincinnati, Ohio. This was the birth of bar coding.

No standard code existed in 1967 but many had been proposed. The criteria for a coding system was clear. Any code must be easy to communicate to a computer. Normal numeric and alphanumeric characters were not foolproof enough, and they could only be read from one direction. The code also had to be readable from any direction because goods could not be oriented before scanning. There simply was no time to orient goods scanned. In addition, it had to fit on small items as well as large and had to be printable on at least 25 different materials from cardboard to cellophane. It took until 1973 for the Universal Product Code to develop.

This UPC code was an almost instant success from its announcement in April of 1973. By 1975, over half the items in grocery stores had the UPC code on the label, and by 1980, the estimate was 90 percent. The grocery industry estimates that over 80 percent of all grocery stores will have scanner checkouts by 1987. The benefits are enormous and it has proven to be the most cost effective way to handle products at the checkout register for the following reasons:

It improves accuracy.

It speeds the input of data.

It updates inventory records immediately.

It assists material handling.

It improves productivity.

BAR CODING SYSTEMS

Although the most familiar to Americans, the UPC symbol is not the only coding system in wide use today. The various systems have different uses and thus different characteristics. The UPC or Universal Product Code is a numeric, binary, fixed length, variable size coding system. The length is six or twelve digits having four elements per character (two bars and two spaces). This code is read omnidirectionally so it has taken hold in the paperback book, magazine, and grocery industry.

The Europeans have their equivalent to the UPC in the EAN or European Article Number. This code is the same as the UPC except that it is longer to identify the country of article origin.

Code 39 was developed for the department of defense and the auto industry to handle alphanumeric data. This code reads from either end (bidirectional) but isn't suited to the omnidirectional scanner. It consists of nine elements per character (five bars and four spaces). Its major advantage is in its ability to code letters, special characters, and numbers.

The Interleaved 2 of 5 code was developed as a bidirectional reading numeric code having five elements per character (two wide, three narrow) with the characters interleaved. The auto industry uses this system.

Federal Express uses the CODABAR system as does the Library of Congress, pharmaceutical manufacturers, and the photographic industry. CODABAR handles both numeric and special characters and uses six alpha characters as a start and stop pattern. It reads bidirectionally and has seven elements per character (four bars and three spaces). This system has each character independent so length is infinitely variable. This code is less likely to be misread, thus its success in pharmaceuticals where errors are intolerable.

INDUSTRIAL BAR CODE USES

Bar codes in manufacturing are not limited to finished goods labeling but extend throughout the factory. They are being used in the following ways:

1. Process monitoring
2. Assembly instruction display
3. Assembly verification
4. Part status reporting
5. Shipping bill of lading generation
6. Receiving bill of lading verification
7. Warehousing location and part identification
8. Automatic routing or sorting
9. Cycle counting
10. Historical record automation

A computer-controlled process gets instructions from the bar-coded part number which sequences operations to make the part. Quality in that kind of operation is automatic.

A bar-coded part read in an assembly area coupled with a computer terminal displays assembly sequences and graphics pictures to instruct the assembler in proper assembly. Parts scanned after assembly verify that the assembly is correct. Each part moving through manufacturing can be monitored for its position in the process to obtain fast part location on the factory floor.

At shipping, the bar codes of the materials loaded on the truck are read and the data used to produce the bill of lading for the shipment. At receiving, the bill of lading is compared to the products actually received by using scanned codes. The warehouse location (bin number) and part stored there is verified or tracked easily by using bar codes.

Assembly lines can be sorted into multiple lanes from a single supply source by reading the part codes and shifting gates to the appropriate line. Cycle counting parts in the warehouse is simplified and more accurate by reading the bar codes on parts. This reduces duplications and provides easy rechecking of off-count inventories.

Bar codes are also used to provide information on exactly which part, having what serial number, was shipped or made on what day at what hour for development of product history records.

Bar codes are simply the best available technique to get accurate information quickly on any part or operation. Their use will become more widespread with time. They are not a substitute solution for errors generated by long part numbers but are a means to more quickly handle short ones. Read error rates drop for bar coded parts but part number verification by computer record check is still necessary to verify that the number read really exists. The shorter the number, the fewer read errors for machine or man.

8

Bills of Material
and the Manufacturing Link

George Plossl said, "After all, the engineers create the bill so that, by definition, somebody other than the designer can make the product. The bill of materials is, therefore, really made for others in the first place. And it would seem to follow that it should be structured for the user's, not the designer's convenience." It is the designer's job to supply users with a useful bill and to do this the designer must know how the bill is to be used.

The bill of materials is perhaps the most important set of numbers in a manufacturing operation. The bill of materials defines the product, provides the material planning framework, helps determine the sequence of operations needed to produce the product, and assists in scheduling manufacturing operations. This is a far cry from the simple bill of materials or parts list engineers learned to put on the mechanical drawing of the product in engineering school. The role of the bill of materials has changed dramatically in the past 20 years.

The following is a generally accepted list of the uses for a bill of materials today:

1. Product definition
2. Manufacturing assembly parts sequence
3. Engineering change control
4. Service parts control
5. Liability and warranty protection
6. Planning material procurement

  7. Scheduling the manufacturing plant

  8. Order entry option selection

  9. Pick lists

10. Scarce material analysis

12. Product costing

13. Pricing of optioned products

14. Backflushing inventory relief

The traditional product definition role of the bill of materials has been known since man made stone-headed clubs and spears. This is the familiar parts list for the product. In those prehistoric times, the parts needed to make an axe were a stone or metal head, a handle, and something to hold the handle and head together. The number of parts has changed little for the axe in half a million years, although the materials may be somewhat different. Today the handle may be hickory, ash, steel, or fiberglass; the head, steel; and adhesives and/or wedges hold it together. Nevertheless, something has changed in half a million years. Our cave-dwelling ancestors made an axe when they needed one and only one at a time. Today mass production makes axes by the hundreds.

We begin by listing the types of bills of materials:

1. Parts list

2. Family tree

3. Indented

4. Single-level

5. Multi-level

6. Summarized

7. Batch

8. Phantom and pseudo

Figure 8-1 shows an engineering parts list. This is simply a list of the parts contained in the final product. There is no assembly sequence, subassembly identification, or other manufacturing information contained in the engineering parts list.

The family tree, shown in Figure 8-2, shows the parts and the subassemblies required as well as the composition of the final product. The indented bill of materials uses indentation as a way of differentiating parents from components, all of which are commonly referred to as levels. A level is "a position in the bill of materials; the parent item is at one level, all of its components are at the next lower level."[1]

---

[1]For additional terminology see George W. Plossl, *The Glossary of Manufacturing Control Terms,* 2nd ed., 1985.

Figure 8-1. Hoist Engineering Bill of Materials

HOIST

ENGINEERING BILL OF MATERIALS

Part No.	Description	Qty. Req'd.
37000	Hoist	1
23822	Motor	1
00842	Drum	1
09387	Drive Shaft	1
01385	Hook	1
13873	Cable	16ft.
51011	Gearbox	1
83996	Electrical Cable	8ft.
71052	Control Pendant	1

(Courtesy of George Plossl Educational Services, Inc.)

Single-level bills of material are those with a single parent and its components only. Figure 8-3 shows a single-level bill and the indented bill is multi-level. The multi-level format shows all parts at all levels.

Summarized bills are essentially the same as an engineering parts list, but they group like items and total quantities of items that are duplicated. These bills are useful for determining the total component quantities required and also to design engineering for standardizing parts usage.

Batch bills are used in the food processing and chemical industries. The most familiar batch bill is the recipe for a cake. The reason a batch bill is different from a conventional bill is that quantities are batch-size dependent. The ingredients in a recipe for one cake are not simply doubled for a cake twice as large. The taste and texture are simply not the same so two batches must be made or the recipe changed to make one twice as large. Two of the most common ingredients needed in batch quantities are salt and yeast. In general, doubling the recipe increases the salt by 1.5 times. Yeast for fermentation of wine or beer is not proportional to batch size either. Applications such as these use batch bills.

Pseudo and phantom bills are bills that describe products or collections of items which do not really exist. A phantom bill of materials is a bill containing items such as a subassembly made on a feeder line and consumed at the assembly line rate. This item physically exists but isn't stocked or scheduled in the inventory planning system. During normal explosions of material

Figure 8-2. Hoist Manufacturing Bill of Materials

HOIST

MANUFACTURING BILL OF MATERIALS

Part No.	Description	Qty. Req'd.
37000	Hoist	1
23822	Motor	1
63141	Drum Assembly	1
00842	Drum	1
09387	Drive Shaft	1
00100	Shaft Blank	1
01385	Hook	1
13873	Cable	16ft.
67343	Gearbox Assembly	1
51011	Gearbox	1
61399	Control Pendant Assembly	1
71052	Control Pendant	1
83996	Electrical Cable	8ft.

(Courtesy of George Plossl Educational Services, Inc.)

requirements, the phantom items just "blow through" to the components because no inventory exists for the item and lead times for its production are nil.

A pseudo bill of materials is a grouping of items that are not actually made but simplify master scheduling and bill maintenance. A grouping of all hardware common to a product line is useful for planning the requirements

Figure 8-3. Hoist Single Level Bill of Materials

Number	Item	Qty Req'd
37000	Hoist	1
23822	Motor	1
63141	Drum Assembly	1
67343	Gearbox Assembly	1

for kits of parts, although kits are never actually assembled. The inventory system does not recognize these items for the same reasons it doesn't recognize phantoms—zero inventory and lead time. The use of these pseudo items is effective in planning requirements for these common parts.

THE THREE SUBDIVISIONS OF BILLS OF MATERIAL

The structure of the bill of materials for the product depends upon the intended use. In design engineering, the bill of materials is a parts list. The list needs to contain all the parts required to make the final finished item. Dimensional drawings define each part; stock item part descriptions are substitutes for drawings for commercial fasteners and similar common items. This engineering parts list contains the quantities of each item required to completely assemble the finished item. In the case of end items assembled from modules, the modules are on the finished part drawings, and the module drawings contain the parts required by each module. The engineering parts list is the design engineering bill of materials—the first subdivision.

The second subdivision is the manufacturing bill of materials. The requirements of manufacturing for subassembly identification, semi-finished components, and assembly sequence information adds data to the parts list. These data include additional part numbers for semi-finished parts and for subassemblies not defined by design engineering, so that drawings may not exist for these parts. Lack of parts drawings doesn't create problems in manufacturing, but it may for design engineering if the rule prevails that part numbers require drawings. The attempt to make drawings for each part number issued seems rational for finished parts in the design parts list but may be unnecessary for subassembly definition. Semi-finished parts are a convenience for manufacturing and they don't need drawings either.

In the early days of production and inventory control, parts were ordered by the order point technique. The order point technique attempted to predict the usage of parts on an item-by-item basis and maintained a buffer inventory level called safety stock. The concept was to order an item in economic lots where orders were triggered by reaching the safety stock inventory level. Usage predictions for the item then defined the safety stock inventory level that would hopefully prevent running out of the item. This method of inventory control requires no information about bills of material. Modern systems using Material Requirements Planning techniques, however, use the requirements for items in the master schedule to plan the requirements for all the components. This means that the bill of materials must be accurate. The question now becomes, What should be master scheduled? It is possible to master schedule finished products in some cases, but for highly optioned products it is frequently impossible to forecast or master schedule the finished part. This results in structuring the bill of materials for planning in such a way that forecasting and master scheduling are effective. Structuring is arranging the bill of materials data into a form that is effective for material planning and manufacturing. It has nothing to do with the arrangement of the data within a computer database file.

PLANNING BILLS OF MATERIAL

The third subdivision arranges the bill structure to make a planning bill of materials. The planning bill improves planning and control activities required to make the product. This bill contains the parts in the engineering parts list along with the manufacturing information in the manufacturing bill. These data are structured to create a bill of materials to simplify the planning process. The structuring process frequently requires additional part numbers (drawings not required) for collections of common parts or planned modules composed of parts with percentage quantities per assembly. Because the planning bill is so important to modern manufacturing control systems, an adequate understanding of the structuring process and the various structures has become just as vital to engineering in a manufacturing operation.

The planning bill of materials is used to obtain the components necessary to meet the planned production in the master schedule. This bill is structured and designed to obtain the parts needed to produce assemblies in the quantities needed to meet the production plans just in time. This is fully compatible with the concept of just-in-time manufacturing and is necessary for adequate production planning.

Planning bills are structured in different ways. The following is a short list of the types of structures frequently used.

1. Modular bills
2. Percentage bills
3. Inverted bills
4. Family bills
5. Super bills

Perhaps the easiest way to examine these different structures is by example. We will use the common overhead hoist to provide our product model. The hoist is diagrammed in Figure 8-4. Note that the common hoist has a total of 2400 possible assemblies (30 motors times ten drums times four gear boxes times two control pendants times one hook). The company wants to make 50 hoists per week. The plant has to determine which 50 of the 2400 possible they will make each week for the next 15 weeks to obtain the parts for an unstructured bill of materials. An unstructured bill of materials usually results in just such impossible forecasting problems and cannot be tolerated if companies wish to maintain minimum planned inventory levels.

Figure 8-4. Hoist Arrangement

HOIST ARRANGEMENT

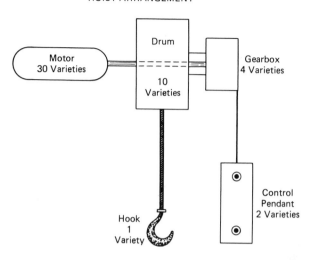

(Courtesy of George Plossl Educational Services, Inc.)

The solution to the forecasting problem is to see the hoist as an assembly of components as shown in Figure 8-5.

One can forecast at the end item level but if the forecasting were to be done at the module level, it would simplify the forecasting problem significantly. It is easy to see that it takes 50 control pendants per week, and now the problem is to determine how many of each of the two kinds to plan. The same is true for the other modules. In total, this modularization of the product

Figure 8-5. Forecasting Options

FORECASTING OPTIONS

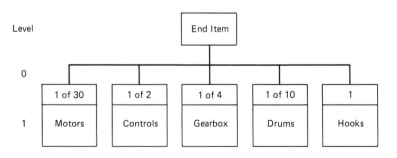

(Courtesy of George Plossl Educational Services, Inc.)

components reduces the number of forecast items to 47 ($30+10+4+2+1$). This makes modules easy to master schedule.

This concept of modularization can be carried further to simplify planning. Consider the control pendant for a moment. The control pendant consists of a case casting, a cover, pushbuttons, a cable and cable connector, and hardware to hold it all together. Group all the parts common to control pendants and issue the group a part number. This item really exists only on paper, but these common parts are needed 50 per week to meet the requirements to build hoists. Give the unique parts in each of the two hoist pendants part numbers also. These parts bags don't really exist either. Apply the same process to the other modules until a series of common and unique parts bags exists on paper. Each bag has a part number, and the quantity required is 50 per week for the common parts in each module. This completes the production of the modular bill of materials structure. The engineering parts list has been structured to simplify component acquisition. Figure 8-6 diagrams this modular structure.

Further extension of this concept produces a single part number for all the common parts in all the hoist modules. This is the "supercommon" bill of materials. This structure has a definite problem, however. When engineering changes are introduced in hoist designs, the supercommon "bag of parts" item must be changed if parts commonality changes. The computer "where used" reports are frequently useless for finding these planned but not constructed, supercommon items. Bill maintenance is therefore a bigger problem but planning is much easier. Engineering must be aware if supercommon bills are being used and be careful to inform planning as soon as bill changes are contemplated that may affect these bills. Figure 8-7 shows a supercommon bill of materials for the hoist.

Percentage bills of material are useful in companies with a fairly stable mix of items and a great many options. This type of bill is useful and effective in a book manufacturing company where paper, glue, cover stock, cloth, headbands, thread, wire, and ink make books. There are literally thousands of combinations of these materials for book manufacturing. The book is defined by customer order and item specification. In this environment, the master pro-

Figure 8-6. Hoist Modular Bill of Materials

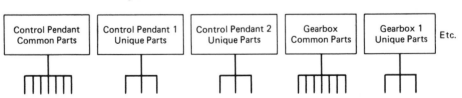

(Courtesy of George Plossl Educational Services, Inc.)

Figure 8-7. Hoist Supercommon Bill of Materials

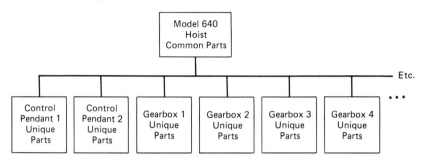

(Courtesy of George Plossl Educational Services, Inc.)

duction schedule for the next two months breaks into two parts. The first three weeks are customer orders, and the next five are forecast demand using percentage bills developed for book families. This permits getting the materials needed to fill customer orders, thus compressing material acquisition lead time. This lets the company quote a shorter manufacturing lead time. Families must be carefully defined to make this work effectively, and widely varying orders within families will produce widely varying percentages in the bills of material. Thus all bills may not be eligible for treatment as percentage bills, but many can use this technique to provide planning information.

Another use for a percentage bill is in the film and pharmaceutical manufacturing industry. Here the problem is packaging. Aspirin packages in 25, 50, 100, 250, 500, and 1000 tablet containers. Film is packaged in one, two, or three roll film packs, in special promotion packages, or supplied with cameras and in bulk reels. In both of these industries the total quantity needed of the product, aspirin or film, can be fairly accurately forecast. The problem is how to forecast the packaging. The conventional bill of materials structure makes the film or the aspirin a component of the final packaging. This presents a problem. If the company wants to overplan the packaging to permit product mix variations to be handled more easily, they will get more aspirin or film than they will really need. A solution is to use a percentage bill in the inverted form. Figure 8-8 shows the standard bill form for film and packaging, and Figure 8-9 shows the inverted form. Note that the total packaging needed is 110 percent of the film. This obtains flexibility in packaging with changes in product mix but retains the overall forecast for the expensive film component. This is a good example of where the percentage bill is useful.

Difficulties abound with percentage bills of material. The bill must be modified every time the sales mix changes to reflect accurate material requirements. If these mix changes are frequent, the variability of material plans will have a ripple effect on vendors and work centers. The second major problem

Figure 8-8. Film Standard Bill Of Materials

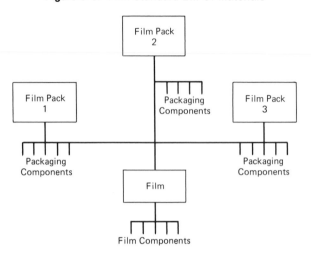

(Courtesy of George Plossl Educational Services, Inc.)

is with seasonal volume changes. In the book industry, the diary presents the biggest problem. The company plans are to make them all year at about a constant rate, but they sell at a high rate from July through January, which presents difficulties in using a percentage bill of materials. The third problem with percentage bills is that future mix changes, known today, are difficult to put into the system. Systems don't usually permit quantity per assembly re-

Figure 8-9. Film Inverted Bill of Materials

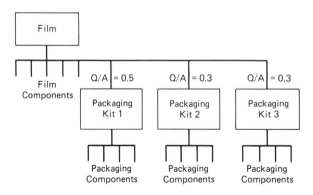

(Courtesy of George Plossl Educational Services, Inc.)

visions by time period when they explode material requirements. There are some software packages that handle quantity per assembly by effectivity date but they are rare.

Engineers need to be aware that percentage bills exist and that inversion to accommodate packaging can be a useful planning tool. They also need to understand that percentage bills will need significant bill of materials maintenance.

Perhaps the planning bill of materials most useful to engineering is the family bill of materials. It describes components needed by a family of end products. This type of bill is useful when a great number of end products is produced from relatively few components. The ceramic dinnerware business that makes dinner plates, cups, saucers, and the like in millions of different styles and decorations finds this type of bill very useful for obtaining raw materials. It is also useful to engineering because it shows how the whole product line relates to the raw materials. This is useful for material standardization studies as well as material control.

The super bill of materials is a variation of the percentage bill. It is a percentage bill of materials for a product family. This term is applied to a variety of finished products that are put into a percentage bill for a product family and is thus useful to companies with a wide variety of end products and no modularity. Again, the aim of structuring is to obtain the data necessary to procure the materials.

COMMON PROBLEMS AND UNIQUE SOLUTIONS

Perhaps the most common problem for materials control is the one caused by engineering creating the comparative or "same as except" bill. This type of bill is often referred to as an ADD/DELETE bill. Essentially, if product A is made using components B, C, D, E, and F, it is easy to get product A1 using B, C, −D, E, F, and G. This, in effect, substitutes a G for a D to get the new variation A1. From an engineering point of view this is fine. The shop could even build it this way, so what's the problem?

The problem is in material planning. Figure 8-10 shows a bill for an automobile that has a dashboard with a hole in it for a clock. The automobile comes with or without a clock. A blank plate is provided to cover the hole for the no-clock option. The clock option adds a clock and deletes a blank plate. This causes no difficulty in engineering or manufacturing, but sales cannot forecast exactly how many cars will have clocks. Sales thinks that 60 percent of the cars will have clocks but they want to overplan for the clock option. They want to forecast 65 percent clocks to have more clocks than they need because of forecast error. Here is the problem. If the real orders received are 60 percent clocks but the plan is for 65 percent clocks, the clock option will have deleted 5 percent of the blank plates needed to cover the holes.

Figure 8-10. Add/Delete (Comparative) BOM

ADD/DELETE (COMPARATIVE) BoM

ADD - ON ONLY BoM

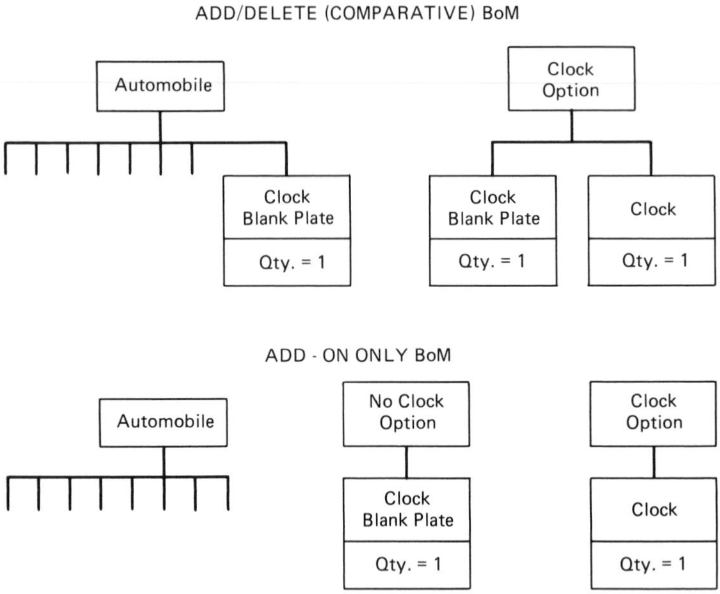

(Courtesy of George Plossl Educational Services, Inc.)

The solutions are several in number. Order 5 percent more blank plates into stock on an order point basis, ship cars without covers over the holes, put in clocks at 65 percent whether ordered or not, or fix the bill of materials to eliminate the problem. The bill is easy to fix by simply using an add only system. For a clock, add a clock. For no clock, add a cover. Regardless of the business, add/delete bills of material are a problem to material planning. They must be eliminated and engineering must understand why.

A common problem for engineering comes about when tools are consumed by the process. Tools need sharpening or refurbishing periodically. On the surface this problem may seem minor, but tooling always seems to be out for repair when it is needed most. The problem may be phrased as, How to obtain the product using tooling which must be repaired and do so with as little upset as possible. A similar problem that fits into the same category concerns expendable grinding wheels used in manufacturing. If the grinding wheels are primarily associated with one or a small number of products, the need for wheels links with the need for products. In both these cases, the manufacture of the product itself calls for the expendable tool or the wheel. If a product calls for a component, the component belongs in the bill of materials.

One solution is to put the expendable tooling into the bill of materials.

Figure 8-11. Bearing 2341 Expendable Tooling Bill

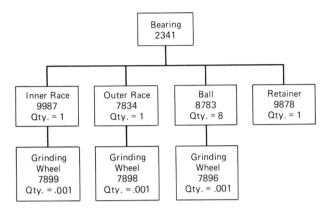

(Courtesy of George Plossl Educational Services, Inc.)

The difficulty is to determine the fraction of the expendable tool built into the product. Now requirements for the product call for a quantity of tooling to meet the tooling needs. Material requirements planning handles the tooling as any other of the components. This has proven to be very useful in many companies. Figure 8-11 is an example of how expendable grinding wheels were specified in a bearing company's bill of materials.

Engineering must understand the planning bills for essentially two reasons:

1. Engineering should avoid problems such as those found with add/delete bills, and
2. Engineering should make use of techniques which improve the ability to standardize parts and plan expendable tooling.

This means that engineering's home should be on the shop floor where the manufactured product needs to flow smoothly and continuously through production. Since bills of material start in engineering with the parts list and become the starting point for material procurement and production, engineering is involved and has a primary responsibility.

THE MANUFACTURING LINK

To get a clear picture of the importance of bills of material today, the operation of modern manufacturing control systems must be understood. The master production schedule is the detailed schedule for quantities of products, described by the bill of materials, to be produced during a specific time period.

An example of a master production schedule is shown in Figure 2-4. It is the planning bill of materials and the structures of the bills of material that convert that master production schedule to detailed order schedules. The better the stability of the master schedule, the better the plan for materials will be. The structures used in the planning bills of material stabilize or disrupt the master schedule. This is where material procurement starts and engineering has a major effect on the ability of a company to effectively master schedule.

Material Requirements Planning is elegantly simple. Figure 8-12 shows a material explosion for an end product. The value and importance of the bill of materials can be clearly seen.

The master schedule is shown at the top; the plan is to produce 20 units in week 2, 30 units in week 4, 25 units in week 6, and 35 units in week 8. To produce 20 units in week 2, 20 units of final assembly components will be required. We have ten now and will get ten in week 2 so the need for final assembly components has been met. The 30 final assembly components required in week 4 are not yet covered. If it takes one week to get them, then an order must be placed in week 3, shown by the "start" quantity. This order to make 30 in week 3 generates a requirement for the subassembly's components in week 3. An order has already been placed for 30 and delivery is expected in week 3. This satisfies the need for this component to make the 30 units of finished product in week 4.

Now jump to the 35 planned in week 8 of the master schedule. The requirements for final assembly components have not been satisfied so the need is generated in week 8 and the order placed in week 7. This produces a requirement for 35 subassembly components in week 7. This need has not been covered and calls for 35 in week 7. With a two-week lead time, the need for subassembly components starts in week 5. This produces a requirement for 35 in the purchased component in week 5 as well. An additional 10 units of this purchased component is required each week for spare parts, giving a total requirement in week 5 of 45. The needs for this purchased component for weeks 1 through 4 can be met by the 90 in stock, but the requirement for 45 in week 5 will produce a need for 20 more. With a four-week lead time and an order quantity of 100, an order must be started for 100 in week 1. Thus, the order placed in week 1 for the purchased component is generated by the planned needs for the final assemblies eight weeks out. That is the logic of MRP and the forward visibility it provides.

The modern system truly links engineering to the entire manufacturing process. The link is no longer limited to fragmentable operations such as parts lists, routings, standards, methods, and the like. Engineering is linked to manufacturing and top management through the planning bill structure. Through material requirements planning, the bill also has a direct effect on the ability of the company to effectively plan and control inventory and production. Modern manufacturing control systems put engineering in a pivotal role in manufacturing companies.

Figure 8-12. Material Requirements Planning

Master Production Schedule

Week No.		1	2	3	4	5	6	7	8	9
Will Make			20		30		25		35	

Final Assembly Component

		1	2	3	4	5	6	7	8	9
Require			20		30		25		35	
Have Now	10									
Will Get			10							
Need					30		25		35	
Start				30		25		35		

Subassembly Component

		1	2	3	4	5	6	7	8	9
Require					30		25		35	
Have Now										
Will Get					30					
Need							25		35	
Start					25		35			

Purchased Component

		1	2	3	4	5	6	7	8	9
Require		10	10	35	10	45	10	10	10	10
Have Now	90									
Will Get										
Need						20	10	10	10	10
Start		100								

(Courtesy of George Plossl Educational Services, Inc.)

ENGINEERING CHANGE CONTROL

The causes for change in manufactured items are many and varied. These causes group into two categories based upon timing. The first group has changes made immediately. These are the mandatory changes. Some of the reasons for changes in this group are:

Product failures
Government regulation
Liability minimization
Material shortage

Product field failures can result in the need for immediate redesign of the product. Few companies can survive in a competitive marketplace with a design that fails prematurely. For example, a company produced valves. It made a ball valve assembly which was used by homeowners to connect two hoses to one hose spigot. This "Y" contained two ball valves set in a split plastic socket which was held tightly to the ball and in place by a press fit brass spacer. This design was easy to make and the market was good. After a short product introductory period, the company began receiving letters from customers complaining of failures. Customers were complaining that the brass spacer worked loose and the ball valve no longer shut off flow completely. This failure occurred where only one hose at a time is connected. The design was checked to be sure a quality problem had not occurred and then the design was modified to prevent these failures.

Government regulation is also responsible for many engineering changes. The most common example is in the regulation of the automobile industry. Recalls have been ordered so car owners can have design defects repaired. There is no doubt that government regulation is a factor in engineering change timing.

Product liability is another area where timing of changes is critical. If a product fails in service and injures someone, that failure can result in heavy damage suits. Companies who find failures which open them to product liability have an incentive to change the design as quickly as possible. The company wants as few flawed products as possible in the field to minimize liability.

Material shortage or unavailability is another reason for immediate engineering changes. Occasionally a supplier will discontinue an item or a raw material will not be available. This results in inability to make a product containing the discontinued item or the raw material. Engineering must redesign the product so parts can be obtained or another raw material used. There is little doubt about timing for any of these changes; they must be made immediately.

PHASED-IN OR OPTIONAL CHANGES

The changes that seem to be the most difficult to make are those that don't have to be made immediately because the timing can be critical. A manufacturer making automobile parts can be asked to provide preproduction samples from production tooling. To make the changed samples the existing tooling must be modified (it's too expensive to make a whole new set of tools) and the present production must stop. The hazards here are:

1. Will the customer order more than expected before the new prototype is approved and released?
2. Will the prototype be accepted as is or will more changes need to be made, further delaying new parts production?
3. Will material volumes in stock and the purchasing pipeline be consumed before they are eliminated by a change?
4. Should new part numbers be issued for the new prototype or should the new part be the same number?
5. Will the old part be required for service?

The first problem is serious because unilateral action by the supplier that shuts down an auto assembly line can result in loss of the account and possible legal action. The best solution to this problem is close coordination between the supplier and the customer. When changes are anticipated, both parties need to be aware of the consequences of all the alternatives before choosing any particular alternative.

With regard to the second hazard, there is no way to know for sure if further changes will be required, but this problem should be weighed with the first in consultation with the customer.

The third question can be answered by accessing the manufacturing control system data. First, material specifications can be checked to verify whether materials on hand are usable. If the new part can be made from the old specified material, the question is moot. If the material cannot be used, the quantity on hand and on order can be checked for any particular schedule change date. This will define the cost in obsolete material and will define when the new material must be ordered to make the new parts on schedule.

The fourth problem of part numbering is more difficult to answer. The basic reason for specifying a new part number is for a change in form, fit, or function. If the new part looks about like the old one, performs the same function, and will assemble with the same mating parts, a new part number may not be required. If the old part must be maintained for service parts or both the old and new parts can exist in inventory simultaneously, a new part number will be needed. The basic question is: Must there be a way to identify old parts and new parts at the same time? If the old parts will not continue to exist when the new are produced and the customer will not readily recognize the change, the change will not need to result in a new part number. If the old parts will exist along with the new and the customer will accept them as interchangeable, a new number is not needed. If the customer will not accept old parts once new ones are produced and both will be available during some transition period, a new number must be issued to keep the new parts separated from the old.

The last question requires a policy on how long service parts will be provided. Some CRT manufacturers offer to replace old CRTs with new ones, at a nominal fee, rather than be forced to carry PC boards virtually forever.

The fundamental rule for issuing new part numbers is to issue a new number if:

1. Old and new parts must be separated in inventory
2. Part form changes
3. The fit to mating parts changes
4. Part function is altered by the new parts

CHANGE REVIEWS

Engineering changes need to be reviewed by all groups who will be affected by the new scheduling. All departments need to know why the part is being changed and the impact that change will have on their operations. All should be consulted on the timing of the changes to minimize unexpected problems. An effective way of handling these change reviews is to use a change review committee. It is composed of members representing marketing, sales, engineering, accounting, materials, purchasing, manufacturing, and distribution. The scheduling of changes is reviewed to be sure nothing has been overlooked and that the needs of all have been balanced. This smooths out the implementation of changes and provides the cross-departmental communications conducive to teamwork. Once passed by the committee, scheduled changes are entered into the computer manufacturing control system.

EFFECTIVITY

The most common system for handling changes in a computer is based on effectivity date. Computers handle dates internally, and thus it is easy to get the computer to make a change effective on a particular date. Essentially, effectivity is the concept shown in Figure 8-13. Part A is presently made from Parts B and C. The new part D will replace part B, and part A will then be made from parts C and D. The effectivity date is the date when this will take place. The greatest difficulty with effectivity dates is that they occur independently of inventory balances, work-in-process levels, and manufacturing realities. Dates are easily measured but in actuality effectivity often needs to occur when materials actually run out—before a new order group is processed or new materials become available. It is rare that these factors occur on any spe-

Figure 8-13. Effectivity

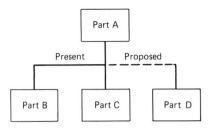

(Courtesy of George Plossl Educational Services, Inc.)

cific date and rarer when a specific date can be predicted on which they will occur.

The solution to this problem is to use effectivity quantity instead of effectivity date. The difficulty with this is that only a small number of changes will be based on runout quantities and computer systems are rarely able to accept both effectivity, date and effectivity quantity at the same time for different parts.

BLOCK CHANGES

One system to schedule changes is very useful, however. This system is the block change. Block changes are used to schedule all changes at the same time. This results in predictable change dates and a single point to minimize the material and plant disruptions that occur. The most notable example of the use of block changes is in the automobile industry. Here changes are made for a model year and almost all major changes are instituted simultaneously. Block changes have some benefits as shown in Figure 8-14.

Figure 8-14. Block Change Benefits

BLOCK CHANGE BENEFITS

- Stable Bills of Material
- Accurate Maintenance
- Accurate Costing
- Minimal Manufacturing Disturbance
- Simpler Service Parts Support and
  History Maintenance
- Better Planning
  Tooling
  Processing
  Phase-Out

(Courtesy of George Plossl Educational Services, Inc.)

Perhaps the best solution to change effectivity is found in the process industry. Here changes are effective when the mix is changed at the inlet. This can occur at any scheduled batch change. If we apply this principle to batch discrete manufacturing, we can schedule changes during a line changeover between models. If the model runs are short, changes are easy to schedule. In their motorcycle operation, Kawasaki makes over 6000 changes annually, but since they manufacture in lots of five of a given model, they don't find change scheduling to be a problem. Changes occur after each lot is completed. With support parts to final assembly made in the same five lots, changes occur smoothly and on short, predictable schedules. Thus engineering changes are best managed where run quantities are short and material flow is smooth and predictable.

9

PROJECT PLANNING AND CONTROL

DEFINING THE PROJECT

The first step in managing any design/development program is the development of a definition of what the program must accomplish. This definition process is essential to obtain a clear picture of the scope and goals of the project. The design/development program is composed of four phases.

The first phase is *preproduction,* which includes those activities in a project that occur prior to actual manufacturing. Research and development occur in this phase. R & D in manufacturing companies includes pure research, but research is usually directed toward production of specific products or processes applicable to the manufacture of products. This is more commonly called applied research.

Application designs are directed toward specific markets and customers. For high-technology industries, application designs include specialized circuit designs, experimental microcircuit designs, and other state-of-the-art designs for specific applications. This is application of existing technology to meet specific product needs.

Although methods, standards, and tooling are manufacturing engineering responsibilities, product development projects frequently require development of new manufacturing processes. Methods design, standards development, and tooling designs for new processes are part of preproduction.

Unproven new processes and equipment are often required to produce new products. The availability of new process technologies frequently make new products possible. When new processes and equipment are required, acquisition, installation, and debugging are part of preproduction. Needs for additional pieces of equipment to increase existing manufacturing capacity are specified during preproduction. The execution phase includes purchase, installation, and troubleshooting of existing equipment types. This is subject to acquisition lead times for equipment, however. If the lead times for obtaining equipment for capacity increase are long enough, they will be purchased before the start of the execution phase.

Many experimental activities occur during preproduction. As a general rule, each must be complete before production can commence. Project phases don't start and stop abruptly, but they overlap as other activities do. Experimental activities normally start first, but experiments may continue far into the production phase. Preproduction is the longest and least predictable phase because of its experimental activities. This phase is the most difficult to schedule and complete on the projected end date.

OPERATIONS PLANNING AND CONTROL PHASE

Operations planning and control starts with the decision to enter a design into production. This decision triggers completion of a general sales, marketing, financial, and business plan to produce the product. This business plan is used to develop a production plan called a Master Production Schedule (MPS). The Master Production Schedule plans production quantities by time period over some scheduling horizon. This MPS is verified by rough-cut capacity planning to be sure average capacities in plant work centers and in vendors' plants are adequate to support the plan. Once verified, the Master Production Schedule drives Material Requirements Planning (MRP) and detailed Capacity Requirements Planning (CRP).

This phase is frequently overlooked by engineering because it is formulated by the production planning group. Many decisions made during the design process will affect operations planning and control. For example, the bill of materials for the product will be used to detail the material requirements for subassemblies and semi-finished parts required by designs. The design also dictates some of the processes needed to produce the product. This defines the work flow and work center and vendor capacities required. Maintenance and machine mean-time-between-failures (MTBF) helps determine how reliable the manufacturing processes will be. The preventive and breakdown maintenance programs may need adjustment to accommodate increasing capacity utilization from new designs.

A thorough understanding of the relationships between product design, process selection, and operations planning and control is essential to project success. Successful projects for manufacturing are those that meet technical, sales, marketing, manufacturing, financial, and plant utilization plans. Incorporation of operations planning and control considerations into a project is essential to smooth manufacturing. Both engineering and manufacturing must work cooperatively on new designs for new products to meet this criteria.

EXECUTION PHASE

The execution phase tests the plans and the project planning system. This phase is either pure agony, with missed budgets, delays, and unforeseen problems, or a pure delight with activities coming together on or near schedule and budget. The agony is inversely proportional to planning care and competence. This does not mean that a good plan insures success, but a well-executed plan prevents failures from causes not related directly to experimental activities.

FOLLOW-UP PHASE

The follow-up or review is the last phase of a project. Reviewing projects improves the ability to plan and execute future projects. Unfortunately, there is usually little time made available for this phase or it simply becomes a fault-finding session. Project reviews are conducted to determine what problems occurred, how they were handled, and what can be done to avoid similar problems in the future, not to find someone to blame. Completing projects successfully requires skilled planning and problem solving. Improving these skills is the primary goal of the project review. The follow-up phase also includes efforts to resolve any lingering production problems. Since these problems may be the result of earlier decisions or activities, they provide vital insights for avoiding similar difficulties in subsequent projects.

PROJECT DEFINITION

What exactly is the project to accomplish? There is a vast difference between designing a product which functions to meet a customer's needs and designing a product to be produced efficiently using existing manufacturing equipment, processes, and materials at a saleable price. Although new product designs should be producible in the current plant, this is not a primary objective for most designers. Rigorous, careful definition of the project is vital to success. The project definition objectives must be met during execution or the entire project can fail. Most experienced managers have seen projects that have been technical successes but business and financial failures. Failure to define the objectives is a frequent cause of this.

The final project definition should include the following parts:

1. Product functional goals
2. Product departmental requirements
3. Management and business goals

The product functional goals include a detailed technical definition of the product. This definition includes the desired technical product specifications and preliminary conceptual module list for modules in the product. These specifications must be realistically achievable. It is counterproductive to incorporate impossible specifications or tasks into a project. Tasks and specifications that are not theoretically achievable must be eliminated. Speculative elements such as anticipated breakthroughs, however, are included. When speculative tasks appear, the difference between desired and mandatory goals must be clearly defined. It must be clear if a certain specification or task is desirable and a breakthrough is required to achieve it, or if the whole project hinges on satisfactory completion of that task or attainment of the specification. Desirable features are flexible, mandatory ones are not. Special attention must be devoted to activities whose outcomes are mandatory.

Many company departments are directly or indirectly influenced by projects. Departmental requirements planning defines what is each department's contribution to budgets, inventory costs, schedules, material and labor productivity, and other resource needs. The documents containing the defined departmental requirements constitute an organizational impact statement. Carefully analyze and discuss the organizational impact to insure departmental support for the project as well as planned implementation at the department level.

The final part of project definition is the overall business plan impact statement. It specifies sales and marketing goals, capital consumption goals, cash flow requirements, customer service goals, and productivity targets. It handles top management's concerns about the project's impact on the overall business goals.

Project plans are designed to achieve project definition targets. While the targets are not normally fixed, changes are tolerable only within limits before plan revision must occur. The project planner and/or manager must be thoroughly familiar with the definition of the project and with those target tolerances. The project plan must also include flexibility to accommodate target changes and establish tolerances to define acceptable ranges beyond which the plan must be completely reworked. Tolerance limits help decide when replanning is necessary.

PLAN DEVELOPMENT

Once the project definition exists, the tasks which must be performed are identified. Some iteration between task determination and project definition occurs, but a project definition needs to exist before detailing a plan. The plan must cover all facets of the definition. Departments included and others influenced must provide plan information to the project manager or planner. In this respect, product planning is little different from the planning needed by a professional football team when a new play is developed. Everyone on the team must understand their role and the role of the other players.

The next step is to identify the tasks required to complete the project. The project technical requirements must be understood by the planner to make the task list. Planners don't have to know how to perform each task but they must understand the resources required for each task. For example, it is sufficient for a planner to know that a new energy control computer requires a new circuit board design and that the new board design will require a circuit diagram, layout, and tooling. The planner need not be a circuit designer.

An information flow diagram is an essential part of the early efforts to split the project into identifiable tasks. It defines where information for each task originates and where it will be disseminated. Flow diagrams provide information about the organizational interactions required to complete the project. They trace both information and material flows and they describe department interests and interrelationships. For example, the office services manager may get a copy of the personnel list sorted by department for mail route planning. Projects that impact the movement of intra-plant mail or that shift personnel within the plant are vital to this manager. Although a product development project doesn't directly affect the office services manager, he or she will need personnel movement information. Without a thorough investigation of informational flow, data routing for projects may be insufficient to enable smooth execution.

Once the major project tasks have been listed, the subtasks required to complete each major task are identified. This step-by-step refinement produces a complete and comprehensive list of tasks essentially by moving from larger chunks to ever smaller ones. It also assists the planner in identifying unfamiliar tasks.

Once developed, the task list is reviewed to determine if the tasks will satisfy the project definition. It is vital that those tasks which directly impact business, management, or technical goals be identified for special consideration. These tasks become the ones to receive the greatest scrutiny and are very apt to be the ones management uses to evaluate the success of the entire project. Early identification and review of these tasks are especially beneficial.

TASK SEQUENCING

The terms *event* and *activity* are helpful in discussions of project management. An activity consumes resources and takes time. An event is an instant in time. The initiation and completion of individual tasks or activities are events. Projects are interrelated networks of activities and events.

GANTT AND BAR CHARTS

The simplest way to plan and schedule a project is with Gantt or bar charts. Activities are represented by rectangles on a two-dimensional graph. Each activity is allocated a single unit on the vertical or Y axis. The length of each bar (rectangle) corresponds to the time required to complete the activity. Each activity rectangle is displaced to the right an amount of time corresponding to the time lapse between project initiation and the task start time. This approach is useful to schedule small projects or for projects that are simple variations of past ones so that the activities are familiar to all concerned. Large projects or projects where unknowns abound are not good candidates for this planning and scheduling technique.

Some tasks can be done simultaneously or in parallel with others; some must be done sequentially. Sequence tasks in the order in which they must be completed. The bar or Gantt chart was primarily designed to control the time element in a project and helps little with sequencing. The Gantt chart planning system is not well suited to complex design projects because:

1. Planning and scheduling are shown together.
2. They are too simple to detect critical slippages.
3. They don't show dependencies well.
4. They are difficult to keep up to date.

NETWORK DIAGRAMS OR BUBBLE CHARTS

The network diagram or bubble chart was developed in the mid-fifties to overcome these Gantt chart deficiencies. A network diagram shows graphically the events and activities needed to complete the project, along with the sequential

Figure 9-1. Project Network Diagram

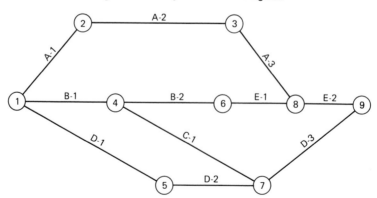

(Courtesy of George Plossl Educational Services, Inc.)

and parallel interrelationships among them. Circles represent events and connecting lines represent activities. The common name *bubble chart* came from the circular representation of events. Dashed lines depict dependent relationships—one event must occur before the next—but activities are not performed between two dashed line-connected events. This network diagram works for computerized or manual scheduling techniques.

The network diagram provides a graphic image of the project as a whole and demonstrates the relationships between events in a way no other representation can. Figure 9-1 shows a simple project network diagram.

NETWORK DIAGRAM DEVELOPMENT

Network diagram development is a relatively simple process. All project network diagrams start with an event labeled "start" on the left end and end with one labeled "finished" on the right. Each task is assigned a starting and ending event. A list is made in starting event sequence. This list of tasks and events provides the data to construct the network diagram. All tasks with the same starting event become activity lines drawn between the starting and ending events. Starting events which start many tasks are "burst" events. End events which terminate several tasks are "merge" events. Network diagrams are produced from left to right in the order or sequence necessary to conclude with the end event on the right.

The network diagram is not complete until dependent, activity unrelated, tasks have been shown. When one task must be done prior to starting another but no activity exists, a simple dependency exists. For example, in the construction of a house, the wallboard cannot be put on the inside walls until the insulation is in, the plumbing and electrical wiring is installed, and the outside siding has been put on. This relationship is clearly defined with starting and ending events but the wiring can be completed at several different times. The electrical wiring may be done before the outside sheathing is put on or after the insulation has been installed, but it must be done before the wallboard is put up. Wiring is independent of the carpentry tasks and may start anytime after the frame is up but must be done prior to putting up wallboard. Dependent relationships of this kind have dotted line tasks called "dummy" tasks added to identify critical event order.

PROJECT SCHEDULING TECHNIQUES

After the list of tasks, events, and activities is complete and a network diagram has been made, the project is scheduled. Although it is desirable to know or establish the overall project completion date at the beginning, as part of the definition, the schedule for the project depends on data about events and activities. There are many techniques for project scheduling, but further discussion is limited to usable and well known ones.

Every technique requires that the activities and events be identified and that an estimate exist of the time required to complete each activity. These activity time estimates must be achievable, on average, for the overall completion date to be realistic. The estimates, like all forecasts, are always wrong. If they are at least two numbers and they are realistic, the project can usually be completed on schedule and within budget. The plan is achievable, at least theoretically, and effective execution and project control have a chance of delivering the completion on schedule.

PROGRAM EVALUATION AND REVIEW TECHNIQUE (PERT)

A common planning and scheduling technique is the Program Evaluation and Review Technique or PERT. This scheduling technique is particularly well suited to developmental projects where estimates of task times are uncertain. This technique was developed to handle design/development projects.

The PERT technique was developed by the United States Navy Special Projects Office, Lockheed Aircraft Corporation, and the consulting firm of Booz, Allen, and Hamilton. The technique was applied to defense projects during World War II where overall completion time was critical and where there was a great deal of uncertainty about the length of time required to complete individual activities. Because of its original use, PERT is especially applicable to determining the probability of completing design/development projects on schedule.

To apply PERT, all activities and events must be identified and each activity must have a defined starting and terminating event. Events which complete several activities are merge events, and events that start several activities are burst events. All activities entering a merge event must finish before subsequent activities can take place. Burst events start multiple parallel activities, but all may not start at the same instant in time. They must, however, be ready to start on the burst event scheduled date.

PERT requires an elapsed time forecast for each activity. The forecast must consist of three estimates: an optimistic, a most likely, and a pessimistic. PERT uses the three time estimates for each task to compute the relative likelihood of project completion for a scheduled overall completion date. PERT also identifies the critical path; this directs management attention to the most critical activities.

PERT analysis starts with a project network diagram and adds length of time estimates for each activity. A typical PERT network diagram is shown in Figure 9-2.

Many commercial software packages are available for doing PERT anal-

Figure 9-2. Typical PERT Network Diagram

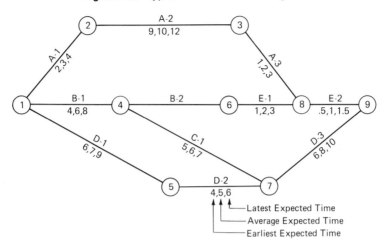

(Courtesy of George Plossl Educational Services, Inc.)

ysis. PERT software exists for almost all mainframe and minicomputers. Most personal computers, with more than 32K bytes of memory, can run commercial PERT programs capable of handling projects with 200 to 500 activities. There are PERT programs for computers with at least 48K bytes of memory whose project size is limited only by disk space available. If a hard disk is available, the maximum number of activities can exceed 60,000. Projects requiring more than 5,000 activities are not usually scheduled on microcomputers because the sheer size of the project produces a need for budgeting, multi-departmental computer support, and a large mainframe to coordinate all the requirements. For large projects, it is advantageous for a project manager to have access to a microcomputer with communications software. This permits the manager to communicate with a mainframe to obtain and to transmit data back on a project. The ability to do PERT analysis, budgeting, and the like on large projects with a microcomputer and send the data back to the mainframe enhances productivity.

CRITICAL PATH METHOD (CPM)

Critical Path Method or CPM analysis is for projects where activity times and costs are accurately known. CPM permits varying resources to compress activity schedules. Compressing scheduled time by addition of resources is called "crashing."

From a managerial standpoint, critical path time and cost tradeoffs are very important. The CPM technique assumes that the activity times and cost estimates are accurate. This is not the case for experimental projects. Although it is very hard to crash the time it takes to invent a nonexistent process, parts of an entire project can be selectively crashed to accommodate errors in forecasting experimental activities. CPM facilitates the identification of what to do to keep a project on schedule at the minimum cost.

The fundamental difference between PERT and CPM is that CPM requires good estimates of the time and costs for project activities under normal and crash (expedited) conditions. Crash time cost estimates are greater because greater resources are needed to execute activities in a crash or expedited manner. Crashing activities requires extra manpower, material, and labor to shorten or compress task completion time. The crash cost estimates permit evaluating various schedule compression schemes for various increases in cost. Earlier completion times increase project costs as do missed schedules. Changes in the critical path resulting from crashing individual activities are easily identified with the CPM technique. CPM provides the costs associated with getting back on schedule when project delays are encountered. Clearly, this is a powerful management tool for design projects where new product introduction dates are critical.

PERT, CPM, AND MATERIAL REQUIREMENTS PLANNING TECHNIQUES

Material Requirements Planning is a technique that nets material requirements against available inventory. It also attempts to determine when orders for materials must be released to get them in time. Although developed for manufacturing products, MRP logic can be applied to projects to smooth resource acquisition and project execution. MRP logic nets estimated project manpower and other resource requirements against available resources. This method determines when orders for resources must be released to get them when needed. Combining the techniques of CPM or PERT with MRP logic yields useful management information and forward visibility in project planning.

One major PERT and CPM shortcoming is that they concentrate on total activity costs and times without regard to the lead times required to obtain resources. Both PERT and CPM assume resources will be available when needed. Any project manager knows that resources are not routinely available when needed and that resource shuffling is normal during project execution. Another shortcoming of PERT and CPM is that these techniques produce no on-hand balance data files to show resource allocation. MRP logic overcomes this limitation by providing automatic on-hand balance files and issuing order launch dates offset by resource acquisition lead times. Applying MRP logic to a project in conjunction with PERT or CPM will minimize overall project cost and improve resource scheduling. This technique is Project Resource Requirements Planning (PRRP).

Project Resource Requirements Planning or PRRP combines the properties of the Critical Path Method or PERT with material requirements planning logic. It yields a time-phased schedule for all material, machine, and human resource requirements needed by the project.

PRRP requires estimates of the quantities of normal and crash resources required to accomplish each task, as well as the overall project and critical path schedules in the form of a bill of materials. In addition, item records need resource acquisition lead times. The resources required for each task comprise a bill of needed resources, analogous to the bill of materials in manufacturing. The project schedule end-date defines when the project tasks end. This is the master project schedule date, comparable to the manufacturing master production schedule. The project master schedule quantity is one (1), because project completions occur only once.

Applying MRP logic, we subtract the resources required by time period from the resources available (netting) to arrive at time phased net requirements for each resource. With this forward resource visibility, resources can be obtained in time to meet the project master schedule date. Even in the case of selectively crashing activities, analysis for available resources is a reality. Project management now develops resource plans for various schedule contingencies. Figure 9-3 shows the typical bubble chart of Figure 9-2 converted to a

Figure 9-3. Tree Structured Bill of Resources

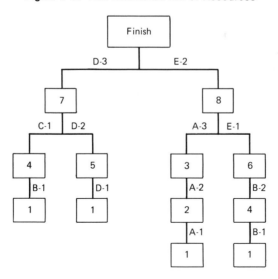

(Courtesy of George Plossl Educational Services, Inc.)

Figure 9-4. Single Level Bill of Resources

SINGLE LEVEL BILL OF RESOURCES
FOR ACTIVITY B-1

EVENT 4

Drafting (time)
Secretarial Services (time)
Sr. Engineer (time)
Jr. Engineer (time)
Technicial (time)
Model Maker (time)
Machinest (time)
Steel (material)
Aluminum (material)
Screws (material)
.
.
.
.

(Courtesy of George Plossl Educational Services, Inc.)

tree structure for MRP analysis. Figure 9-4 shows a sample bill of resources required for the tree in Figure 9-3. The resource/project chart shown in Figure 9-5 clearly identifies when resources are needed and in what quantity.

DEVELOPING THE PROJECT SCHEDULE

Project schedules and plans are based on forecasts or estimates of the time each task consumes and the resources required. Project schedule review for cumulative critical resource requirements is a form of rough-cut capacity planning. Capacity planning validates the schedule. The schedule is workable if,

Figure 9-5. Resource/Project Chart

TYPICAL RESOURCE PROJECT CHART
RESOURCE REQUIREMENTS PLAN

Resource No. R65   Descr.   Sr. Eng.   On Hand 1120   U/M - Hrs.   Safety Stock 100
Order Qty. L4L   Lead Time 10   Alloc. 0   Action - None

Week No. P/D	1	2	3	4	5	6	7	8	9	10	11	12	13	14	15	16
Require	40		90	10	10	40		90	40		90	10	25		30	10
Sched. Receipt																
Available	980	980	890	880	870	830	830	740	700	700	610	600	575	575	545	535
Plan Due																
Plan Released																

Pegged: Project P165
              Project P384
              Project P113

on average, the total resources are available. If the resources are not available on average, the project schedule cannot be achieved. Cost estimates based on assumed resource availability are useless; the resources must be available. If the resources are not available but can be obtained in time, the resources are really available. It is essential to analyze projects to determine if resource requirements are realistic. This validates the schedule. The PRRP technique is invaluable for this purpose.

Even though the total resource requirements are available on average, the project must be reviewed for detailed resource requirements in each time period. Although the project may not exceed the resource capacity available in total, projects often require peak resource levels beyond those available. Resource shortfalls on critical path activities cause project delay. CPM and PERT analysis shows the critical path of the project but neglects resource consumption. PRRP techniques augment the PERT and CPM techniques by providing resource consumption data. Since the project manager is inevitably swamped with problems even in well-planned projects, time simply isn't available to manually replan the project during execution, and computerized PERT and CPM are actually not much help in saving time at this stage. If PRRP is used, critical path resource needs and capacity shortfalls can be included as part of the original project plan.

Project planning truth number one is that Murphy never takes a vacation; something will go wrong. If the project is well planned and contingency plans exist, the overall project schedule can be met by selectively crashing some tasks on the critical path. Schedules based on fully crashed critical paths are schedules that fail. Keeping a project on schedule is directly related to managers' ability to anticipate problems and react. They must react before problems delay the tasks on the critical path. The forward visibility and the ability to selectively crash activities improve performance in meeting the project schedule. PRRP, PERT, and CPM all contribute to effective schedules.

UPDATING THE SCHEDULE

During project execution, progress is closely monitored. The schedule must be updated because:

1. Management wants progress reports.
2. Resources need to be allocated to tasks to keep the project on schedule.

The need for management reports is secondary to the need for project control. This doesn't minimize the need to provide management with information, to monitor progress, and to coordinate management activities, however. The primary requirement for effective control is timely allocation of resources. Some tasks get ahead of schedule and resources are reassigned to others that are behind, smoothing execution. Periodic project replanning permits recognition of and reaction to such opportunities. The project manager periodically reviews progress, reallocates resources as necessary, and updates the schedule. This permits timely actions to keep the project on schedule. PRRP is simple to update. Activities can be added, deleted, or modified by changing or adding records to the bill of resources. Usage records are updated by resource usage transactions. The computer runs or regenerates a new plan whenever one is required. The logic is identical to that used by the business running a manufacturing control system.

The project manager establishes a contact person in each task group who will provide feedback on the present status of task completion. Although completion percentages are usually estimated, not calculated or measured rigorously, the manager weighs the information fed back to determine task status. Task personnel tend to be optimistic about schedules; managers must compensate for this condition. Project managers must be technically competent to discuss and evaluate task status. Their understanding of the technical details enables them to recognize when an individual activity is far enough behind schedule to jeopardize the entire project. This is especially important in projects having crashed schedules or those having high visibility. Timely and accurate feedback often identify actions needed to get the task back on schedule.

Feedback must be forthcoming from all tasks in order to exercise activity control. Budgetary importance, visibility, complexity, and whether or not the activity is on the critical path determine feedback frequency. Feedback is necessary for nontechnical support activities as well as for the technical ones. Coherent execution of the whole project is impossible without it.

If PRRP helps plan a project, resource consumption is automatic. Records update as resources are consumed, and the computer determines balances. PERT and CPM techniques don't have this record-keeping capability. Planned versus actual resource consumption is essential and automatic for PRRP. Better resource consumption information means better project management.

Many inexperienced project managers measure percentage of resources consumed and equate this to the percentage of completed project. Resources consumed are not indicative of elapsed time or percent complete. Redirection of resources from one task to another results in lower individual task resource consumption but little, if any, change in project completion percentage. When activities are being crashed, monitoring elapsed time is more useful than monitoring resource consumption. Crashing activities to return to schedule increases resource consumption. For creative activities, resources consumed are not indicative of completion or difficulty of the task. New equipment requires "up front" expenditures of financial resources with little project progress. Time is the only true measure of schedule performance on individual tasks. The experienced project manager tracks the overall schedule by monitoring actual event occurrences against plan.

CRASHING CRITICAL ACTIVITIES

There are three activity crashing philosophies.

1. Crash early critical activities to provide a cushion against later difficulties.
2. Crash activities only when the project schedule has slipped on the critical path.
3. Crash every activity to get the project done as early as possible.

The third is impractical. The cushion provided by the first approach is attractive but produces heavy front-end financial consumption. Experienced project managers use the second approach to get back on schedule. Since smooth execution results in better budget performance and fewer stress-related problems among personnel, experienced managers crash as few tasks as possible.

Noncritical activities provide project resource allocation flexibility. These activities are selectively delayed or advanced to stay on schedule. Resources are borrowed from noncritical activities and used to crash critical ones. Completing noncritical activities early wastes resources. Good project management is allocation of adequate resources where and when required to complete all tasks on schedule. Resource balance and smooth project execution indicate good management—not early completion, brinksmanship, or full-time crisis operation.

EXCEPTION CONDITIONS

Project management is management by exception as much as management by objectives. The objectives are the project plans. Managing the exceptions keeps projects on schedule. Progress monitoring determines where project tasks are in trouble—hopefully the exceptional case. Successful project management is timely action to control and eliminate exceptions.

PROJECT BUDGETING

Based on preliminary estimates, top management approves projects. They approve the projected costs and benefits. Those estimates are forecasts. The first rule of forecasting is that the results will be wrong. Good estimation requires forecast error control. Project budgeting attempts to do that. Budgets are developed for each task in the project and the results are totaled. Although budget consumption is not an appropriate measure of completion of an individual task, it provides an early indication of major deviations from cost estimates. Such information is extremely valuable.

Once a workable project schedule has been developed, clear-cut task definitions are available to use in budgeting. Developing task budgets consists of identifying the resources needed to complete the task within the allotted time and estimating their costs. The costs for specific resource categories accumulate, yielding total resource budgets. Inconsistencies between these budgets and the preliminary cost estimates must be resolved before the project proceeds. This step helps eliminate management surprises which undermine a manager's credibility.

HUMAN RESOURCE BUDGETS

Listing the different types of human resources needed to complete each task is the first step in budgeting labor costs. For each labor category, the time (man-hours, weeks, months, years) is estimated. These totals are then multiplied by the labor rates to develop a budget for each category. Once these costs have been developed for each task, it is easy to accumulate labor costs for the total project.

An extensive technical background and a familiarity with the design/development process is essential in project budgeting. Technical expertise is required to translate design task definitions into estimates of the types and numbers of scientists, engineers, designers, and technicians needed for each task.

The kinds of human resources needed for development and testing range from machine operators and cost estimators to door-to-door product demonstrators. Pay particular attention to skill identification during human resource budgeting. Failure to identify the necessary skills results in serious estimate errors.

Labor needs during commercialization include manufacturing engineers, industrial engineers, and plant engineers. Design engineering incorporates changes that facilitate the manufacturing process. Familiarity with manufacturing processes is essential for those who determine human resource needs for commercialization of a design/development project.

THE MANUFACTURING-PROJECT MANAGEMENT LINK

Planning, scheduling, and controlling ongoing manufacturing operations is both similar to and different from project planning and scheduling. Both require that certain activities and the resources necessary to support them be carefully planned, scheduled, and controlled. They are different in that project control concerns one-time execution. Manufacturing executes the manufacturing plan and produces the product repetitively. Companies produce multiple products simultaneously and repetitively. Thus manufacturing must manage hundreds of products at the same time.

On-time production dictates that the right materials be available in the right quantities at the right time in the right place. Complex multi-level products consist of hundreds of parts. Production requires dozens of tools, fixtures and jigs, and similar supplies. The absence of any of these items at a critical point will delay production just as delays on critical path items delay projects.

The PRRP technique is directly extended into manufacturing if manufacturing is using a manufacturing control system with MRP. This means the entire project can be integrated from the start event on the network diagram through manufacturing and distribution. This project planning system permits an integrated view of the project for both engineering and manufacturing. The logic is common, so both engineering and manufacturing understand the project schedule, and the project can be integrated through manufacturing from the start event on the engineering network diagram. The end event for the project becomes the start event for manufacturing. The plan is clear from engineering and valid schedules are possible from new product introduction dates back through all engineering activities. This is by far the most effective tool to control new product introductions in manufacturing.

10

LEAD TIME AND SETUP TIME

THE MAKEUP OF LEAD TIME

Lead time is the length of time it takes to complete a task from the time the task starts. Lead time is composed of several parts: ordering time, queue time, setup time, and task execution time. For purchasing, lead time is the time between ordering and material receipt. In a manufacturing operation, lead time for production of a given product is measured from the time the shop order is released to production until the work is completed. The components of lead time in that environment are well known. Figure 10-1 is a lead time bar chart which shows graphically the makeup of lead time.

The largest fraction of lead time is queue time. This is the time the prod-

Figure 10-1. Lead Time Bar Chart

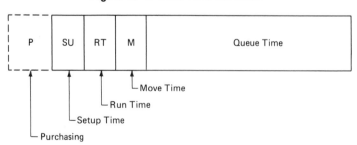

(Courtesy of George Plossl Educational Services, Inc.)

uct sits on the plant floor. It sits waiting for other components to be received from the warehouse or vendors, for the work center to be ready to work on it, and for all the paperwork required for the job to be obtained.

THE EFFECT OF ECONOMIC ORDER QUANTITY ON LEAD TIME

Another fraction of the total lead time is setup time. Setup time is the time required to change over the equipment needed to make different batches. This component of lead time is quite critical and has been studied for many years. The importance of the setup time component relates to its use in the formula for square root economic order quantity (EOQ). This formula balances the cost of ordering against the cost of carrying inventory. The cost of equipment setup is a large fraction of the ordering cost used to calculate the EOQ. The greater the setup time, the greater the EOQ and the greater the lead time.

For example, let's suppose we have an operation which makes zinc die castings. The normal setup time is six hours, and the EOQ calculates the economic order quantity to be 5000 pieces. The total run time for the lot size is 5000 times 30 seconds per piece, or 41.6 hours. The total of setup and run time is 47.6 hours. This is approximately one week of production on one shift per order. This company will probably ask its customers to order capacity in weekly shifts by setting the minimum lot size at 5000.

If we reduce setup time by one quarter to 1.5 hours, we need to recalculate run time, setup time, and lead time. The EOQ changes to 2500 pieces (square root of 1/4 times the old EOQ). The run time is the same per piece so the batch or lot run time is 20.8 hours. The total setup and run time is 22.3 hours. The lead time minimum is 22.3 hours. Removal of 75 percent of the setup time has reduced the lot quantity processing time to 46 percent of its original level. This has now cut the order quantity in capacity units to one half shift per week. Assume a customer was placing orders for 5000 pieces every five weeks. The reduced setup time and resulting lower EOQ permit ordering 2500 pieces every week and a half. This halves the lead time, which improves customer service and smooths the movement of money and material through the manufacturing plant.

What happens if the setup time drops to eight minutes? The EOQ formula yields a quantity of 745 pieces. The run time and setup time for the most economic lot size is now 6.3 hours. Machines produce in lots of one shift per day (8 hours). This permits customers to order in daily increments. The customer that ordered 5000 every five weeks had an average consumption rate of 1000 per week. It is now possible for the plant to supply this customer with weekly deliveries from daily production. This releases capacity to four other customers every week for better customer service. This dramatic reduction in lead time for the customer and in production time in the plant makes it possible to make on Monday and ship on Tuesday and have the product delivered by Friday. Lead time reductions improve customer service and cash flow simultaneously.

Customers can order in small lots of 750 to 1000 parts, and they can be supplied in eight hours of process time. If that same customer orders in lots of 5000, the process time jumps to 48 hours. This is over a fivefold increase in minimum lead time. It is unlikely that the customer uses all 5000 parts as soon as they are received and would probably prefer to order just the quantity needed every week. If the customer really needs 750 a week, customer service improves with lots of 750 run and shipped on a one-week lead time. Cash flow becomes the piece parts cost times 750 per week. The overall assumption in all of these calculations is that the run quantity equals the EOQ. Setup times, although a small part of the total lead time, are a very important factor in lead time length. The batch quantities required to order in economic lots serve customers better when small batches are run on short notice. Customers know better how many parts they really need because of shorter forecasting horizons. A purchasing agent generally has a much better idea of what is needed next week than what is needed 15 weeks from now.

LEAD TIME MANAGEMENT

Lead times drop by cutting batch quantities as a result of reducing setup time. What is not commonly understood is that reduction of work-in-process from smaller batches and fewer jobs on the floor also cuts lead times even more and makes them less variable. A simple simulation demonstrates this well.

QUEUE SIZE—LEAD-TIME SIMULATION

Two work centers with a process time of one day each are set up. This is a perfectly balanced operation. Jobs flow out at the same rate they flow in and are processed. If no jobs are in queue at the start, the lead time would be two days; one day in each work center.

The simulation runs with a number of empty plastic or cardboard containers representing a lot of parts to process. For this simulation, about 40 such containers will be required. Mark 30 of the containers with an X and 10 with the numbers 1 through 10. The numbered containers will permit tracking lead time through the work centers. Job tracking sheets and a summary sheet complete the materials for the simulator. The job tracking sheets are sheets with 30 numbers or numbered squares on them. Figure 10-2 shows a sample tracking sheet. One of these tracking sheets is required for each simulation run. The summary is a sheet such as the one shown in Figure 10-3. It summarizes the lead time for each of the numbered jobs. The first simulation runs

Figure 10-2. Sample Tracking Sheet

1	2	3	4	5	6
7	8	9	10	11	12
13	14	15	16	17	18
19	20	21	22	23	24
25	26	27	28	29	30

Figure 10-3. Job Summary Sheet

Job No.	____Jobs In Queue	____ Jobs in Queue
1		
2		
3		
4		
5		
6		
7		
8		
9		
10		
Total		
Average		

with a queue of two jobs. Put two of the boxes with X's at each work station to complete the simulator setup.

Use one person at each work station and one person to record the boxes on the tracking sheet as they exit the second work center. Release job number one to the first work center on the first day. Random job selection is necessary. Insure that workers choose jobs from the queues without seeing the numbers and without following a pattern. Release job two on day two and so forth for all ten numbered boxes. Flush the numbered jobs from the queues by releasing X's after the tenth day.

Fill out the summary sheet by subtracting the job number from the day on which it completed the second work center. This information is on the tracking sheet. A typical "two jobs in queue" summary appears in Figure 10-4.

Run the same simulation with four jobs in queue and the summary will appear like the one in Figure 10-5.

This simulator demonstrates the truth about lead time and queue size. The larger the queue size, the longer the lead time and the more variable that lead time will be. That is the true relationship between lead time and queue size.

It can be argued that a priority planning system would prioritize jobs and they would not be chosen at random as in this simulation. True, but if the numbered jobs were prioritized to get them through in two days the X'd jobs would then take longer, and the result would still be as demonstrated. The average is four days for two jobs in queue and eight days for four jobs in queue. The longer the queue, the longer and more variable the lead time. That's the real truth.

Figure 10-4. Two Jobs in Queue Summary

Job Number	Lead Time
1	4
2	6
3	9
4	2
5	2
6	3
7	5
8	3
9	4
10	10
Total	48

Average lead time: 4.7 days
Standard deviation: 2.78 days

Figure 10-5. Four Jobs in Queue Summary

Job Number	Lead Time
1	8
2	4
3	7
4	13
5	8
6	16
7	5
8	11
9	6
10	11
Total	89

Average lead time: 8.9 days
Standard deviation: 3.78 days

This simulator also suggests what to do when work centers experience breakdowns. Assume the work centers are operating with two jobs in queue and the second center stops producing because of a machine failure. Two alternatives exist—shut down the first center or continue to run it. The lead time for a given center is the queue size (two) times the process time (one day). Each center produces an average lead time of two days. If we continue to run the first work center and the second is down for two days, the queue at the second center will rise to four jobs. The average lead time will rise to (4×1) four. Center one plus center two now has a lead time of $2+4=6$ days.

If the first work center were shut down when the second one failed, the lead time would stay at an average of four days and two days of production would be lost. All jobs would still be late by two days, but the effect of extending lead times increases lead time variability, which further reduces customer service. Letting the first center run with the second down will cause the jobs to be late by two days plus the increase in lead time variability. This means that it is better, from a customer service viewpoint, to shut down operations rather than allow queues to build. This assumes both can catch up when operations restart.

When sufficient capacity exists, it is possible to let the first operation continue to run and not build queues. If the balance on the first and second operations was achieved with both operating one shift, queues could be worked off by operating two shifts to catch up when the second operation was fixed. Please note, however, that the same result occurs by running both operations on two shifts after completing repairs. The key is to recognize that the queues cannot be allowed to rise if lead times are to be stable. Using overtime to make up lost capacity is cheaper than building queues, considering all effects.

TOOLING CONSIDERATIONS

Lead time is a function of tool design. If tools are designed to permit short setups, then batch sizes will drop with a reduction in setup cost (based on the EOQ formula). Small batches will smooth out material flow and will shorten lead times. Thus, tool designs are directly responsible for a portion of long lead times and large batch sizes. Although more expensive tooling may be required to get short setups, the effect of lead time on batch sizes and work-in-process inventory must be considered. Some tool designs force long setups and short setup tools cost little more.

Consider the case of the zinc die casting operation discussed earlier. What does it really take to get setup time down to eight minutes from six hours? When the setup operations required in the six-hour setup were studied, the longest time element was waiting for the die to get hot. The dies were initially mounted cold in order to accurately align them on the head plate. But when they were mounted hot, setup time was reduced dramatically. This was done by making die block sizes the same for all cavity sizes and aligning the blocks using a pin register system. The dies can be mounted hot because alignment is preset to the hole and slot locations in the die block. Bolting down the die can also be done hot. The die is preheated by placing it on top of the melt furnace a few hours before it is needed.

What else can be done to reduce setup? The shut height of all dies can be made constant so machines no longer need adjustment for this variable. The last major variable left was the volume of the cavity and resulting ram adjustment. Part cavity size is different for every casting so the ram needed adjusting for every cavity. Cavities cannot be identical in volume but dies can be. Tool design computes the cavity volume for every casting and designs cylindrical slugs into the die to equalize die volume. Good engineering has reduced setup time to eight minutes by modification of die designs.

These cost savings must be measured against higher costs for the dies because the material required for the die blocks is no longer matched to cavity size. It is constant. Alignment pins and locating holes must be made and maintained accurately. Cavity volumes must be computed and slugs produced to make volumes constant. All of these add cost, but what are the savings? Savings are much harder to quantify. What are the savings associated with shorter lead times? What is the value of improved customer service from short lead times? How much will market share improve with shorter lead times? These questions are much more difficult to answer quantitatively. Manufacturing and engineering need accounting help. Smooth production flow and short runs which build little work-in-process inventory improve total inventory turns. These improvements in inventory turnover can be predicted with accounting's help.

The route to short lead times is clear. Excess work-in-process must be avoided. Setups must be short to make short runs economic. Tooling must be modular to shorten setups and provide flexibility. Time-consuming manual adjustment of dies and other tools must be eliminated as much as possible.

Group technology is a key to short lead times and is discussed in detail in Chapter 11, "Manufacturing Engineering." Another key to short lead times is having manufacturing capacity available when needed. Available capacity requires equipment which runs reliably. Equipment must be kept running. The key to reliable equipment is effective preventive and breakdown maintenance. Plant engineering and maintenance is responsible for this function. The plant engineering role is discussed fully in Chapter 12, "Plant Engineering."

THE TRUTHS ABOUT LEAD TIME

Lead time is what you say it is.

Lead time is what you make it.

Lead times increase with increased work-in-process.

Lead time increases as EOQs increase.

Lead time is what a company says it is. If a company tells customers it takes 14 weeks to produce products, it receives 14 weeks of orders. If the company tells its planning and control system to use 14 weeks to plan to make products which have a 3.5 day total process time, it takes 14 weeks plus variability to produce the products. As long as a company doesn't tell its planning system that the lead time is less than the process time (3.5 days in this case), any smaller planning lead time can be used. The smaller the planned lead time, the lower the variability and the better the customer service.

Relatively constant order backlogs indicate adequate capacity within the plant to make the total product volume. This means that the inflow of orders and outflow of shipments balance. If balance exists, no job reservoir or backlog is needed. The flow is fluid and continuous; the next job follows the one previously processed. Jobs are processed in the order in which they are needed. Backlogs are useful to smooth out fluctuations in capacity required. With a backlog, the problem of which job to process next consumes much management and computer time. On average, capacity is adequate but job priority and sequencing is difficult. The greater the backlog, the greater the priority problem and the longer the lead time.

It has already been shown that low work-in-process produces the shortest possible lead time. Backlogs develop and lead times get longer when orders are accepted beyond the capacity of the plant or work center to process them. Backlogs also develop from equipment failures, material shortages, and other production interruptions.

Lead time is what you make it. Companies that reduce work-in-process reduce setup time, and those that work on effective maintenance of equipment reduce lead time. Sales and marketing keep lead times stable by only selling up to the production capacity of the plant. Lead time is manageable.

THE LEAD-TIME WHIRLPOOL

Extending the lead time to compensate for insufficient capacity doesn't cure capacity problems. Engineering is directly involved in both capacity acquisition and lead-time reduction to eliminate capacity problems. Engineers must understand the lead-time whirlpool in order to prevent a company from being sucked in.

Since a plant has a finite capacity to produce a product, increasing the lead time requires the customer to cover his or her needs over a longer period. This extends the forecasting horizon. The customer will invariably compute the quantity needed over the new horizon and add a hedge for future uncertainty. As lead times are extended, the orders receive ever greater manufacturing capacity. Increasing lead times to compensate for insufficient capacity digs the capacity shortfall hole deeper, faster. Increasing the lead time increases the incoming order rate and increases load on an already overloaded plant.

Lead times can be reduced by decreasing EOQs and work-in-process inventory levels, as well as by managing orders based on capacity. If a company has a few customers whose orders consume a majority of the manufacturing capacity, an allocation system can be used to quote short lead times for orders remaining below a predetermined capacity level. Backlogs are now controlled. *Allocate capacity and lead times shorten.* Orders exceeding the allocation level can be delivered without quoted lead time and based on a realistic delivery date. Establish delivery dates which don't exceed the maximum plant capacity to make product, and backlogs won't grow. This requires management discipline to prevent overselling plant capacity. The rule is to plan for plant capacity increases, not to oversell the product, and to reduce customer service.

It is essential for companies to avoid getting caught in the lead-time whirlpool. When plant capacity is at its limit and shipping dates are missed, the temptation is to increase the lead time. However, if lead times are permitted to extend, the whirlpool will start. More orders will be released covering the longer horizon. That increases loads in plant work centers. The increased work-in-process causes queues to get larger. The larger queues cause lead times to extend and become more erratic. This results in more missed shipping dates. To amateurs tinkering with lead time, it appears that the first modest increase in lead time didn't work so a big jump is made the second time. This whirlpool will continue until capacity is added because business appears to be booming. Let it start and be prepared to be sucked in. This whirlpool is diagrammed in Figure 10-6.

Figure 10-6. The Lead Time Whirlpool

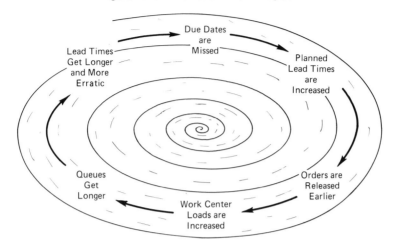

THE SEMICONDUCTOR INDUSTRY—A CASE IN POINT

In 1970 the semiconductor industry suffered a disastrous recession. Sales fell more than 50 percent and shipments declined 25 percent. Most semiconductor companies lost money that year. During the period from 1971 to 1973, major new markets were developed. Consumer electronics, telecommunications, industrial electronics, microcomputers, and electronic automotive ignition systems were born. Industry leaders touted this "pervasiveness" as the way to avoid another 1970 style recession. By 1974, business was booming. Semiconductor suppliers could not keep up with demand. Order backlogs reached record levels, delivery times ballooned to over a year for many chips, and new plants were under construction.

In 1974 disaster struck again. *Business Week,* November 1974, reported "Semiconductors Take a Sudden Plunge." In a few months, the industry's backlog evaporated from order cancellations and returns. *Business Week* drew the obvious conclusion—the semiconductor industry was in a depression. Double ordering was blamed as the cause of this crisis. No mention was made of the real culprit—amateurs tinkering with lead time.

On April 26, 1976, *Business Week* featured the article "A Startling Recovery for Semiconductors." This article noted that prices were firming, delivery times were extending, and some customers feared a return to the capacity crunch of early 1974. Semiconductor companies were urging customers to look further ahead and be more accurate in stating their real needs; that is, they were voicing a plea for better long-range forecasts. This was an impossible dream, and the semiconductor industry planted the seeds of the next recession. Amateurs were still tinkering with lead time.

The events which had led to the recessions were clear and would repeat. George Plossl, in his *News Letter No. 20* to clients, described the real causes.

1. The economy was healthy and semiconductor users were experiencing increasing demand for their products.
2. Semiconductor manufacturers were hard-pressed to meet the increasing demand.
3. Management pressures to improve customer service at the expense of high inventory were common.
4. Customers feared a capacity shortfall and overordered, sometimes doubling or tripling their forecast needs.
5. Semiconductor manufacturers saw backlogs skyrocket and quoted longer lead times for chips.
6. Customers were forced to cover their requirements further into the future. The further forward they looked, the less well they knew what they wanted so they hedged on the high side.
7. More orders were triggered by the longer lead times and order backlogs grew even faster.
8. Chip users increased their own inventories of chips far beyond their real usage needs.
9. The whirlpool had been created and items 5 through 8 repeated until a slight dip in consumer demand for the products with the chips occurred.
10. The bubble burst when shipments finally outpaced incoming order rates. This reduced backlogs and shortened quoted lead times to improve customer service. Coupled with bloated inventories, this triggered order cancellations and the recession.

Increases in capacity were not needed in the 1973–74 period and aggregate demand continued to increase in the 1973–76 period. The boom–bust cycle was preventable.

On July 21, 1980, the *Business Week* headline read, "Rolling With Recession in Semiconductors." After highlighting the dramatic growth in the industry in 1978, 1979, and the first half of 1980, the article stated that the recession had finally come to semiconductors. Heralding the slowdown was a declining ratio of new orders to shipments. George Plossl, in his *News Note No. 49*, stated clearly that this magic ratio did not measure the real cause of the recessions. However, the industry didn't listen to him in 1976 nor solve the real problem. The *Business Week* article went on to say that the slowdown would be "a minor blip on the industry's growth curve."

The headline in the business section of the San Jose *Mercury News,* May

2, 1982, read "Semiconductor Industry Poised for Rebound." This article reported that semiconductors had been "crippled for nearly two long years." The signs of this rebound were unmistakable. Prices were firming, lead times were dramatically stretching out, and orders were outpacing shipments. Semiconductors got caught in the whirlpool again.

BREAKING OUT OF THE WHIRLPOOL

The best approach to getting out of the whirlpool is not to get into it. If in it, stop extending lead times; make them shorter. Shorter lead times are the result of good customer–vendor communications. The event sequence to reduce lead times and break the grip of the whirlpool is:

1. Determine the real production capacity limit.
2. Cancel safety stock replenishment orders.
3. Freeze new order release rates at or below the capacity limit.
4. Smooth out order release.
5. Reduce planned lead time.

Capacity management is key to lead-time reduction. Backlogs develop by putting work into an operation at a rate greater than the rate the operation can complete it. Effective capacity management is as easy to understand as water level control in a bathtub. To let the water (backlog) out of a bathtub (work center), pull the plug (increase output capacity) and/or shut the water faucet (throttle work input rate). When working a bathtub, nearly everyone knows what to do. When working a manufacturing plant, the backlog is measured by load reports while capacity is ignored. Thus lead times extend, increasing the input rate (open the water faucet), and the bathtub fills faster (and management wonders what went wrong!). What's so mysterious?

Load is a measure of the amount of work (volume). Capacity is a measure of work flow rate. The capacity of a work center is the number of work units (standard hours, pieces, etc.) produced per unit time (day, week, etc.). Maximum work center capacity is the maximum rate at which the center can produce product. Actual experience, past history, documents the capacity limits of a work center.

Safety stock provides a cushion against stockouts caused by fluctuating demand. Safety stocks represent stored capacity. When a plant is overloaded, allocating capacity to safety stock items conflicts with allocation of capacity to customer orders. Elimination of safety stock releases capacity to satisfy customer orders. This reduces the backlog created by those orders, shortening lead times.

New order input rates must drop to reduce the backlog. This cannot be done by overselling capacity. Freezing the input rates and forcing them to match or drop below the output capacity prevents backlog growth. This lowers backlogs and shortens lead times.

These measures may appear harsh, but the reason for skyrocketing back-logs and lead times is a lack of capacity management and input-output control. To control backlogs and lead times the rates of work flow must be managed. The solution is clear, but difficult. It is hard work to discipline an unruly plant. There are no substitute cures, only techniques which temporarily block symptoms.

THE MACHINE SHOP CASE

A medium-sized machine shop had 16-week lead times and serviced ten major customers. Two of those customers were other divisions of the same company. The general manager determined that work-in-process could be reduced and lead times shortened. Engineering determined that total processing time on most products took about one week. This meant that 15 weeks of queue time and material was sitting on the plant floor.

It took sales little time to determine that the ten customers amounted to over 90 percent of company sales volume. Production control determined that these customers consumed 97 percent of company capacity utilization. A group technology analysis by engineering defined the six major routings which ac-counted for over 80 percent of the products ordered by these customers. En-gineering further estimated the plant capacity based upon the current work mix.

From the data, the general manager concluded that plant capacity was adequate but marginal. Small additional requirements from other customers were causing overloads in bottleneck work centers. The general manager called two of his major customers. These customers had excess inventory on some items and shortages on others. Armed with this additional information, the general manager developed a plan to cut lead time.

1. Engineering and purchasing would meet with the customers to establish the percent of current plant capacity each consumed.
2. Plant capacity would be allocated to each customer at least equal to that being used currently.
3. Lead times would be cut by one half for orders that fell below or matched that allocation.
4. Orders above the allocation would be assigned free capacity, when avail-able, and realistic delivery dates would be quoted only after receipt of order.
5. Additional capacity would be sought if demand continued to exceed ca-pacity over 30 percent of the time.

Results were dramatic. Customers had shorter lead times (eight weeks) and had an incentive not to overload the vendor's capacity. Customers knew more accurately what they wanted and were better assured of getting it. Because of lower forecast errors, hedge allowances were reduced, freeing capacity for the parts vitally needed. This worked so well that lead time was subsequently cut in half twice more. When lead time fell to two weeks, customers were happy, vendor–customer communications improved, and sales quoted shorter lead times than their competition. This plan worked so well that 30 percent of the plant capacity became available for new customers or to increase allocations available for former customers. Cash flow was better and profits improved.

11

MANUFACTURING ENGINEERING

PRODUCT COMMERCIALIZATION

Well-designed products that perform well in functional testing still may not measure up commercially. It must be possible to sell the product at a marketable price. Clearly, low production costs are important in a product's competitive price. Issues such as process compatibility, product maintainability, and manufacturability must be addressed. Manufacturing engineering has this responsibility. (Refer to Chapter 1 for the functions and roles of engineering.) Management often asks design engineering to change the design to facilitate manufacturing. Design engineering insures these changes don't damage the purpose and function of the original design.

DESIGN FOR MANUFACTURABILITY

Designers choose materials, components, fasteners, and processes based on the ability to meet functional requirements. Designers also generally use the items they are most familiar with. This single purpose—designing the product to perform properly—does not produce the most manufacturable design.

For example, printed circuit boards can be laid out with lengthwise and/or crosswise components. It may be unimportant to the designer responsible for the printed circuit layout which way a particular component is oriented; however, manufacturing does have a preference. If the components are either all lengthwise or all crosswise, automatic insertion equipment does not need to rotate while inserting. A mixture of orientations requires more time, setups, and programming. Additional setups are time-consuming and reduce productivity, so manufacturing likes to avoid them. Although both single- and double-chip oriented layouts work, one is easier to make than the other.

Manufacturable designs are often simple modifications of original designs to suit specific plant equipment. For example, a solid-state local oscillator box had three surfaces that needed machining: the top of the waveguide flange, the top of the box on which the cover sits, and the bottom. It was designed as an investment casting for manufacturing. The designer took the drawings to the machine shop superintendent for comment. As drawn, the design would require the shop to put the boxes on a vacuum chuck to hold them for finish machining. Manufacturing had a problem—no vacuum chuck. They needed bosses on the box to hold it for machining. The superintendent's suggestion was to include clamping bosses as part of the casting design.

In another case, a CRT terminal was redesigned to permit robotic assembly. Robots don't handle fasteners well so the design had to minimize the number of fasteners used. The finished design had a single screw for grounding between the power supply and the main board. Assembly time dropped from several hours for the old design to under 30 seconds for the new. After the successful redesign, it was discovered that robots could not be justified with such a short assembly time. Some side effects were also observed. Manufacturing needed one-tenth the original lead time of older CRT designs. This meant less trouble staying on schedule and getting parts. The plant had less inventory, better flow, fewer parts problems, better quality, and better customer service. This all occurred because the product was redesigned for improved manufacturability.

Good design simplifies the manufacturing processes used to produce the product. Since manufacturers are in business to make products, designers should work directly with manufacturing to produce manufacturable designs. Design for manufacturability must start early in the design/development process. It takes too long to pass the drawings sequentially from design to manufacturing engineering and then to manufacturing. Design, manufacturing engineering, materials procurement, sales, marketing, and manufacturing must be involved from the start. The objective is to design and develop products that can be manufactured and sold most efficiently and profitably.

MODULAR PRODUCT COMPONENTS

One effective way to simplify manufacturing is to use modular components. Assembly of modules to produce final products with many options is a cornerstone of automotive and motorcycle design and production. Designs are conceived for product lines and models designed to fit within those lines. These models use modular component subassemblies. Optional configurations result from modular subassembly selection. The customer receives a custom configured product composed of easy-to-manufacture standard modules.

Some designers believe modularity requirements are excessively restrictive in that they force designers to use standard components and assemblies with standard parts. This restricts designer creativity. Few managers say everything must be standardized, and fewer believe nothing should be. It takes a very creative designer to use standard items to make unique designs.

Modularity is a problem for those designing for one-of-a-kind or job shop production as well. One defense contractor had five vendors for five identical bushings, all with separate part numbers, in five different rockets. Inventory was excessive in all five cases. Further review revealed that all five rockets could have used the same part. This company found that some parts may be standardized for even one-of-a-kind products. The defense contract requirements may require special labeling and a separate part number for each bushing, but the design can still be identical.

The real question is, Can this product design use parts that we already make for other products? The success of efforts to use existing parts is directly proportional to previous efforts to group parts by manufacturing process. Modular products and standard parts are easier to plan, use in manufacturing, and stock in inventory. Modularization must have high priority if products are to be efficiently produced. Unique features can be designed to be added at the last operation, for example, special machining, left versus right hand, plating, painting, or decorating. Although these are design duties, they are included under manufacturing engineering because suggestions of this nature frequently occur there.

ENGINEERING THE BILL OF MATERIALS

From a design/development standpoint, manufacturing engineering cannot afford to wait until design finishes the product drawings to begin its work. Manufacturing engineering must work on tooling, methods, and standards just as soon as tentative parts lists exist. Manufacturability improves through minor design modifications without damaging product integrity if a process of negotiation and compromise among departments begins early. The net result of the project team approach is that it takes longer to get designs finalized. This seems counterproductive, yet, once in the production phase, products move faster, thus shortening the production cycle and reducing inventory. In many companies, getting the product into production requires a 50 percent redesign and takes longer than the original design. Adding redesign time to the original design time makes the team approach much faster and better. "We never have time to do it right but we always find time to do it over," is the hallmark of excessive departmentalization.

5. Material
6. Material handling
7. Setups and tooling
8. Job improvement possibilities
9. Working conditions and hazards
10. Methods

1. It may seem obvious, but each operation should have a purpose. Some operations are the result of previous operations. For example, inspections may be necessary because a previous manufacturing operation produces significant scrap. Operations of this nature can be eliminated by studying the previous operation and dealing with the cause of the problem. If the operation is necessary or cannot be eliminated, determine if improvements are possible by doing it another way. For example, automate the inspection at the previous operation rather than have an inspection operation later.

2. Product design changes may make certain motions in assembly operations unnecessary. It is wise to incorporate these changes if they don't affect product operation. Also determine if the product, as designed, can be assembled automatically or by robot.

3. List all operations required. Determine if operations can be eliminated by changing procedures. Study to see if operations can be combined for smoother material flow. Combining operations reduces the number of levels in the bill of materials, further simplifying planning and control.

4. Inspections are required by internal quality control and by outside customers, such as the Defense Department. Quality control inspects parts for dimensional accuracy, finish, material, and other physical properties. Determine if tolerances and other allowances are necessary. Consider changing previous operations to simplify inspection needs. Use statistical quality control at operations to eliminate the need for subsequent inspections.

5. The correct material and material condition reduces the need for some operations. For example, steel doesn't require annealing if fully annealed steel is purchased from the mill. Material utilization is an important factor also. Determine if the material is used to its best advantage. For example, is excess scrap being handled because the strip steel is too wide and waste is excessive?

6. Material handling and material storage are separate issues. For operations analysis, material flow direction and volume are important. Improve material movement through better floor layouts and use of conveyor transport where possible. Set up progressive assembly operations when possible.

7. Setups and tooling are important for smooth manufacturing. Certain tooling designs improve setup time; others retard it. Is tooling designed to minimize setup; if not, why not? Are setups performed efficiently? These factors all need careful study.

8. Are performance improvements possible from job changes? Can "drop delivery" be employed? Drop delivery directly to a transfer conveyor

METHODS AND STANDARDS

Fred Taylor demonstrated that time standards could be set on industrial w
His work to develop industrial standards was to compare methods of doi.
job. Over the years, time standards have been given heavy emphasis in
dustrial engineering. There are many books on engineering time standards a
methods for setting them. They cover stop watch studies, MTM, PTS, wo
sampling, and other techniques. Statistics have been employed to evaluate tin
studies. However, the primary motivation behind all this activity has shifte
from better methods development to incentive pay computation.

It is widely believed that good standards can improve productivity.
Nevertheless, time standards don't improve productivity any more than book-
keeping makes money for the company. It is the use of standards for better
methods development and incentives for workers that increases productivity.
The methods improvement concepts are worth reviewing because many are
needed to achieve effective material control.

Methods improvement begins with a review of current conditions. Current conditions are:

1. Equipment specifications and design
2. Tools, jigs, and fixture design
3. Plant layout
4. Materials utilized
5. Operations sequences currently employed

Equipment specifications and design influence operations sequences and
motion studies, as do tools, jigs, and fixture design. Plant layouts frequently
dictate material flow direction and workplace layouts. Materials utilized and
operations sequences are keys to further study in operations analysis. Documented, current sequences are required.

OPERATIONS ANALYSIS

This is the procedure used to study operations. It systematically develops opportunities for process improvement. The key elements explored during operations analysis are:

1. Purpose of the operation
2. Product design
3. List of all operations performed
4. Inspection requirements

can substitute for shopbox transfer between operations. Job improvements frequently eliminate subsequent operations. These possibilities must be identified.

9. Working conditions and hazards affect operations in a powerful way. Be sure operators receive adequate training. Look for hazards that cause operators to reduce the pace. Eliminate operator exposure to hazards by employing robots. Painting and welding are two operations in which robots can be used to free up human resources.

10. Methods covers the detailed manual processing sequences performed by workers. Schedule repetitive operations for a detailed time-and-motion study. Determine if automation is possible for repetitive operations.

All these elements aim to smooth the flow of the manufacturing process. None of them addresses engineering time standards for incentive pay. Good time standards are the result of good operations analysis and should be employed for that reason only, not as a punitive measure. Operations analysis improves productivity, performance, and resource utilization.

ROUTINGS

Proper routings are key to smooth material flow and to manufacturing efficiency. For many years, manufacturing engineering has needed to determine the best way to route designs through machine centers composed of similar machines. These functional machine layouts provided flexibility. Jobs and batches could follow any path through machine centers. It has recently been recognized that common routings permit machines to be arranged and operated more like an assembly line or process plant.

Manufacturing engineering is in an ideal position to suggest ways to design the product to use standard routings. Although routings are partially dictated by the part configuration, standard routings can reduce the time required to produce routings for new products. Routing development should take place simultaneously with physical design. This results in compression of the design/development cycle.

PROCESS PLANNING

High-efficiency manufacturing is moving rapidly toward flow processing using dedicated manufacturing cells rather than functionally grouped machine centers. Manufacturing from raw material through assembly is becoming a single, continuous flow process. This approach aims to make batch manufacturing operate more like a process plant. Given the flexibility of rapid setups and the predictability of reliable operations, work-in-process is small and lead times are short when this occurs. The results are high inventory turns, good customer service, and high efficiency.

Process planning at the product design stage is necessary to show how the work will flow in the plant and how that flow will affect manufacturing capacity. Early recognition of capacity problems highlights opportunities to modify designs to smooth manufacturing flow. For example, a solid-state local oscillator for a radar system was originally designed to be machined from solid 6061-T6 aluminum alloy by a numerical control machine. This design would have further overloaded an already overloaded NC machine center. The box was redesigned as a vacuum impregnated investment casting. Impregnation sealed the porous casting (hermeticity was required), and the casting provided a route around the overloaded work center. The boxes still needed machining, but a manual mill could do the flycutting of the surfaces rather than requiring complex NC tapes.

GROUP TECHNOLOGY

Group technology is the arrangement of discrete manufacturing units into material flow processing manufacturing lines. The lines may contain a few or many machines through which parts flow. The goal is to permit manufacture of many parts using similar processes within machine groups or GT cells. For example, many washer-shaped parts require routing to a lathe, a mill, a drill, and a surface grinder. These machines can be grouped to form a manufacturing cell. The cell will process many washer-shaped parts in an assembly-line manner. The only requirement is that the lathe, mill, drill, and surface grinder be used in that sequence. Any part requiring these operations in sequence can use this cell.

IDENTIFICATION OF GROUPS—CODING

For product design, parts must be coded so that they can be identified by types. (See Chapter 7, "Part Numbering and Coding," for details on GT codes.) Screws, bushings, bearings, handles, and shafts are all examples of part types. This coding must be consistent. A bronze cylinder used as bearing surface for a shaft must not be called a bushing by one designer and a bearing by another. Are ball and roller bearings classed with bronze cylinders as bearings? Consistent noun usage is necessary for groups to be effectively coded.

It is wise to code part routings so that similar processing paths can be identified. These codes need to identify particular machines used as well as general processes employed. These routing groups should also include setup requirements for tooling and fixtures. Once routing groups are coded, the process of grouping starts.

GROUPING CODED ITEMS

Group codes facilitate the identification of like parts and the review of similar parts to determine if they can be standardized. Standardization of parts reduces both the number of parts manufacturing must plan to handle and the inventory levels of duplicates.

Coding permits common routings to be found. The process of developing common standard routings begins with a search for matches and near matches. Simply finding common routings does not insure a GT cell is possible, however. Equipment capacity must be effectively used. If the common routing occurs often enough to effectively use the capacity of the machines grouped as a cell, then a cell is practical. The routing codes make part identification easy, so capacity can be determined. If part volumes use the machine capacity effectively, then machines can be grouped to act like an assembly line. The setup requirements coding permits standardized jigs and fixtures to be located or developed. This most effectively reduces the number of parts and simplifies the manufacturing processes used to make products.

Where GT cells already exist, new designs must use the efficiencies of existing cells if possible. New designs should not complicate such cell development. Integration of the product design and processing functions incorporating routings, design, plant machinery, and manufacturing must be exploited for maximum productivity.

TECHNOLOGY COMPARISON— GT VERSUS CONVENTIONAL

Conventional or functional machine layouts can be compared with those of group technology in several areas. Figure 11-1 shows a relative comparison summary of GT and functional layouts.

Investments

The investment in machinery per item produced is higher for group technology for a given capacity level because cells must operate at a speed equal to that of the slowest machine. Since machine speeds vary, a balance at the maximum speed is unlikely. This means that more equipment capacity will be idle or underutilized in the GT layout, and additional equipment will be required to provide equal capacity. In a plant not fully utilizing present equipment, additional equipment is rarely required to create cells.

Inventory is dramatically lower for GT layouts because operations proceed through the cell in small lots of parts, sometimes one at a time. The functional layout almost always completes one operation on the entire lot of

Figure 11-1. Technology Comparison

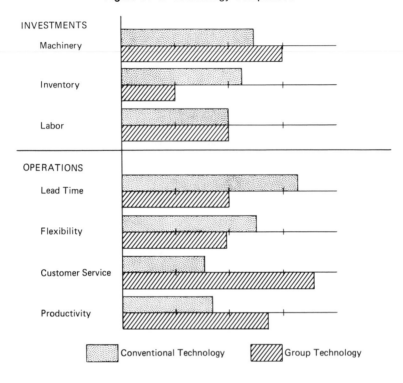

(Courtesy of George Plossl Educational Services, Inc.)

parts before the lot proceeds to the next operation. This produces a slug of inventory flowing through the functional operations and a part-by-part flow in the GT case. GT reduces the inventory at operations. The capital investment of equipment and inventory is usually much lower with cells than with the functional layout.

Labor levels in either the group case or functional case are about equal. Machines still need operators regardless of the layout. The major difference is that in the GT case operators are trained to run any machine in the cell. This means less specialization of operators and more cross-training, but it also means work variety and labor flexibility. This is in contrast to a single operator serving multiple like machines. Machine operators can also serve as material handlers for small parts moving short distances, which reduces indirect labor. The job description for operators is different for GT cells, but the number of workers remains about equal.

Operations

Lead times are significantly lower where group technology is employed. Balanced operations smooth the flow of parts through standardized operations and cause lead times to drop. This is true because work-in-process levels and queues are smaller. This makes the GT layout superior to the functional layout.

Flexibility is lower for the GT case than for functional machine layouts. Anything can be produced by the machines in a functional layout, whereas the GT cell restricts operations to a finite sequence. This limits the flexibility of manufacturing. This is also why it is rare for manufacturing to be completely converted to cells. A combination of cellular and functional layouts is the norm.

Sales and marketing see higher customer service levels on those standard items produced through GT cells. This is because of shorter lead times and greater small lot efficiency. A product produced in a GT cell can be shipped in partially completed lots. The functional layout could do the same only if lots were split into small work order batches but this is rarely done.

Productivity is higher for GT cell processing. Since productivity is output produced per hour worked, the GT cell completes operations at a smooth rate with little or no queue time to delay processing. This improves the output per hour worked.

Figure 11-2 shows a detailed comparison of layout methods on a factor-by-factor basis for the batch process, continuous process, and GT process cases. Figure 11-3 shows a functional layout and a GT layout to produce the same four parts.

PLANT LAYOUT OBJECTIVES

Plant equipment layouts have historically attempted to minimize material handling costs. Colleges and universities offering industrial engineering courses have taught this for many years. The customary process of determining a plant layout has been to measure the material flow volume in pounds, tubs, shop boxes, pallet loads, or forklift trips between work areas. These areas may be work centers, machines, or staging areas. The usual measures of distance employed follow idealized paths composed of straight lines with angular or right angle turns. The material handling costs are then estimated as proportional to the volume multiplied by the distance. Each proposed layout is tested by estimating the volume-distance product, and the layout with the smallest prod-

Figure 11-2. Layout Method Comparison

LAYOUT METHOD COMPARISON

FACTOR	INTERMITTENT	CONTINUOUS	GT
Layout Basis	Functional Similarity	Product Similarity	Part/Operation Sequence Similarity
Flow Pattern	Variable	Fixed	Fixed
Capacity	Unbalanced	Balanced	Balanced
Materials Handling	Manual	Automated	Semiautomated
Dispatching	Work Order Scheduling	Product Run Scheduling	Part Number Scheduling
Flexibility	Much	Little	Some
Work Tasks	Single	Multiple	Multiple
Breakdown Consequences	Limited	Severe	Severe
Feeder Lines	Unsynchronized	Line Driven	Synchronized
Overlapping Operations	Rarely	Yes	Yes
Lead Times	Long and Variable	Short and Fixed	Short and Stable

(Courtesy of George Plossl Educational Services, Inc.)

Figure 11-3. Functional & GT Layout

FUNCTIONAL LAYOUT vs. GROUP LAYOUT

Functional Layout

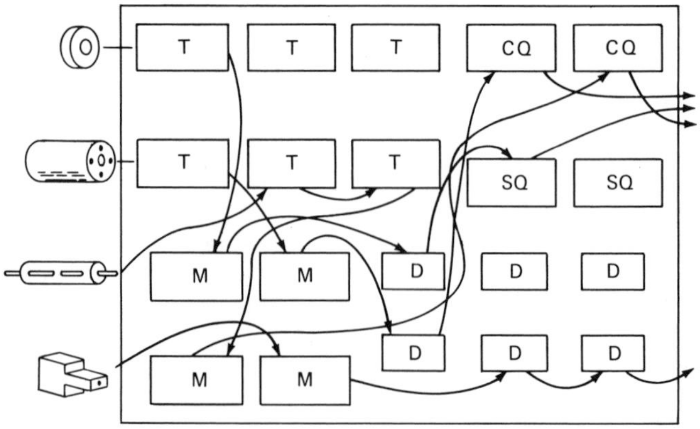

Figure 11-3. (*Continued*)

Group Layout (Flow-Line Cell)

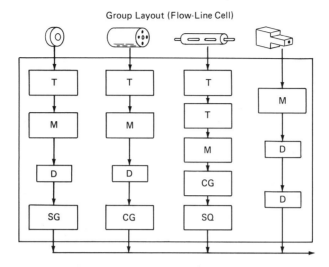

(Courtesy of George Plossl Educational Services, Inc.)

uct is the optimal one. This minimizes material handling cost. An example of this technique is shown in Figure 11-4.

Manufacturing at first concentrated on material volume and distance as cost factors in material handling; inventory was not a factor when the method was developed. Today, however, the emphasis is on inventory levels and short

Figure 11-4. Manual Layout

Link	Flow Volume	Distance	Extension
AB	400	3	1,200
AC	200	4	800
AD	100	5	500
BC	- - -	5	- - -
BD	300	4	1,200
CD	200	3	600

Total [Flow x Distance] 4,300

(Courtesy of George Plossl Educational Services, Inc.)

lead times. The real objective minimizes work-in-process investment, not material handling cost.

The work-in-process investment is not necessarily proportional to product volume. With many complex subassemblies having significant labor cost and other purchased parts needed for the next higher subassembly, the value of parts transported in a common shop box or pallet load may be significantly different.

Thus, the value of the product in the box must be obtained to minimize WIP investment. Since investments are measured over periods of time, the time it takes to move product from one location to another is more important than the distance traveled. It should also be noted that time is not necessarily proportional to distance so distance is not an acceptable substitute. It will be necessary for engineering to work with accounting to get product values for these studies. Figure 11-5 illustrates the change in emphasis from distance and volume to WIP investment and time.

The process of finding the best layout is the aim in either case but the objectives are different. The objective of minimizing the WIP investment directly impacts the cost-time curve for the product being made. The area under that curve is the inventory that is accumulating all along the manufacturing path. Minimizing that area maximizes the inventory turns and thus the flow of money through manufacturing.

Figure 11-5. The Real Objective

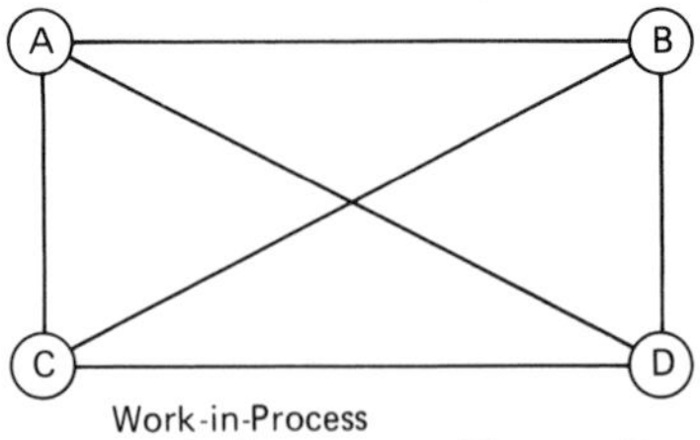

Links	Work-in-Process Dollars	Time	Extension
AB	5,000	8	40,000
AC	10,000	6	60,000
AD	8,000	10	80,000
BC	---	10	---
BD	4,000	6	24,000
CD	12,000	8	96,000

Total [WIP$ x Time]   300,000

(Courtesy of George Plossl Educational Services, Inc.)

COMBINATIONS OF CONTINUOUS
AND INTERMITTENT FLOWS

There are few cases when all material flows can be forced into continuous processes. Batch manufacturing can rarely be converted into a flow like a chemical plant or refinery where material flows in pipes. This forces manufacturing engineering to address production as some combination of continuous process and intermittent discrete machining. The difficulty is in determining what to put in cells and what to retain discrete. There are some basic guidelines to help:

1. Assemble parts in continuous operations.
2. Make cells from balanced high-volume routings.
3. Lay out the plant with operations roughly in line if not in cells.
4. Automate material handling between operations.
5. Avoid backtracking to previous work centers.
6. Build quality in with reliable processes rather than relying on inspection to accommodate deficient processes.

These guidelines will help smooth flow of material and will improve the ability to execute a good production plan. The goal of manufacturing is to produce efficiently only those items customers need as rapidly as possible. The plans developed by any material planning system are designed to permit effective and rapid execution. The job of manufacturing engineering is to determine the most efficient method to use to smooth and speed that execution. This brings up the topic of how to use the output of material requirements planning systems in conjunction with group technology cellular execution.

GT CELL SEQUENCING WITH MRP

The concept of material planning using MRP has one purpose—to get material needed to make higher level assemblies. The quantities of material drop into convenient "need date" time buckets. The order dates drop into time buckets based on the lead time needed to get the components. This means there is some flexibility in the plan, so that if the plan calls for 100 parts in a given week the quantity of 100 and the week are not absolutes. The full lot of 100 parts may not really be needed until the following week and partial quantities may satisfy real demand. This means that short-horizon schedules can be developed for group technology cells.

Short-Horizon Scheduling

Short horizons may be a three-week or one-month period. The period is partially a function of experience and the flexibility of the cell to make short runs. The shorter it is possible to make the run, the greater the ability to schedule execution to make parts in exactly the quantity required to satisfy orders for components in demand at that time. Figure 11-6 shows the MRP-generated requirements for four products produced by a single GT cell.

As long as the total quantity of each of these parts can be produced over the three-week horizon, the actual demands for components can be met. This assumes that the cell can produce parts weekly to meet the needs for components. In effect, this modifies MRP into a three-week time bucket to smooth production flow.

Weekly Balance Cell Run Schedules

The cell, therefore, must produce 500 + 400 + 300 + 300 or 1500 parts over the next three weeks. To smooth the production flow and produce these quantities at an even rate, Figure 11-7 shows the weekly quantities required.

This schedule produces the parts needed in total and achieves a balanced run every week. The question is, In which order should the parts be run? The answer is, In the short horizon schedule. More 7692's and 1276's are scheduled to run the first week than are needed by the weekly MRP-generated plan. The 7106's just match the plan but the 9745's are short. If the 9745's run at the beginning of the week and rerun at the beginning of the second week, the plan will be more closely approached than by any other production sequence. The run schedule is therefore 9745's followed by 7106's followed by 7692's, with 1276's last.

Daily Balance Cell Run Schedules

Let's assume that the cell can be set up on short notice and that the EOQs are small enough to permit daily production of a mix of parts. This means that the cell can produce parts on a daily basis to meet higher level assembly demands. This will reduce overproduction and use of materials before the time when the higher level assemblies really need the parts. This reduces lead times

Figure 11-6. MRP Generated Order Quantities By Week

Part No.	Week No.	1	2	3	Total
7692		100	200	200	500
9745		300	100		400
1276			50	250	300
7106		100		200	300

Figure 11-7. Weekly Part Requirements

Part No.	Fraction	Week No.	1	2	3	Total
7692	500/1500		167	167	167	501
9745	400/1500		133	133	133	399
1276	300/1500		100	100	100	300
7106	300/1500		100	100	100	300

and improves material flow and flexibility. What does the cell need to make every day? Figure 11-8 shows the daily quantities required.

The order in which the parts are needed remains the same as for the weekly schedule—9745's followed by 7106's followed by 7692's, with 1276's last. This satisfies the weekly forecasted demand and the quantities required over the three-week horizon. It gets parts produced in a smoother flow than the lumpy weekly production schedule and provides better flexibility in terms of parts availability to higher level assembly operations. Small batches require less storage and still keep assemblies supplied.

It is useful to note here that when assembly batch quantities match those of the cells and the flow through assembly smooths out, the rate of flow still proves adequate to meet the master schedule for making end items. This schedule also permits assembly to make small lots of end items on a flow basis and permits sales to supply customers needing small daily lots of parts as well as those requiring large chunks. The large chunks can still be made over the three-week horizon, but the capability is there to flow the product at an even rate.

Cycle Balance Run Schedules

As engineering and manufacturing work to shorten setups and smooth work flow through cells, the cell cycles become shorter and shorter. If the cell can cycle in 1/100th of the three-week requirements, the cycle part requirements will work out as shown in Figure 11-9.

With this schedule, a number of interesting things occur. There is no longer a need to make parts in any given sequence. The assembly area receives 6.6 cycles per day of all parts. This is virtually a continuous flow. The assembly

Figure 11-8. Daily Part Requirements Schedule

Part No.	Quantity
7692	33
9745	26
1276	20
7106	20

Figure 11-9. Cycle Part Requirements Schedule

Part No.	Quantity
7692	5
9745	4
1276	3
7106	3

area gets a smooth stream of parts and makes assemblies in small lots. The small lots have dramatic effects on engineering change control because changes can be incorporated at any lot change. Quality also improves because small batches limit the number of bad parts. Part defects are detected earlier and less production interruption occurs. Tooling is easier to monitor, so bad parts are less likely to be produced. Tooling can also be changed frequently for inspection and repair, which further improves quality.

If this scenario appears fictional or impossible to achieve, the facts in real plants prove it can be done. Both Kawasaki and Harley-Davidson produce motorcycles in model lots of five. Parts are produced in cells adjacent to the assembly line, and part production matches motorcycle production. There are also other examples of companies both in the United States and abroad that have adopted these techniques and have lot sizes under ten. This can be done and must be done if pull production is to be achieved in the manufacturing plant.

PULL PRODUCTION SYSTEMS

Pull production draws parts through production centers as needed to fill orders. This is the opposite of the push system, which releases orders to starting operations based on a forecast of what is needed to fill orders.

Perhaps the best known of the pull production systems is the Toyota developed system called Kanban. The concept of pull production is simple: Produce only what is needed to ship to customers. The conventional push system releases orders to production when planned, thus pushing them into manufacturing. The pull system pulls product from final assembly which pulls parts from previous operations. The Kanban card is a device used to implement pull production.

THE KANBAN TECHNIQUE

Kanban cards come in two varieties, a move card and a dispatch card. Figure 11-10 is a diagram of a typical work center. Parts flow from parts storage through the center to finished products storage. In push processing, an order is released to the work center. A batch is made from parts pushed into the work center by the work order. The finished products are pushed into finished storage to complete the work order for this center.

In pull production using the Kanban technique, finished products are pulled from finished product storage. This releases dispatch cards to the work center. The dispatch card tells the work center what to make, and the work center relies on its supply of Kanban move cards to pull parts from parts stor-

Figure 11-10. Typical Workcenter Diagram

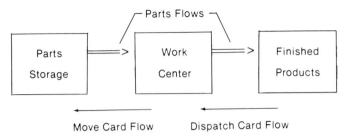

Move Card Flow   Dispatch Card Flow

age. The flow of products is controlled and matched by the flow of Kanban cards. Products are made only when dispatch cards request products, and parts move only when requested by move cards. This card technique controls both the flow of products and the level of work-in-process inventory. Inventory is controlled by limiting the number of Kanban cards available to put on boxes of parts.

Move cards move stock from one location to another. Kanban move cards contain the following data:

1. Part number
2. Container capacity
3. Number of the card (serial number)
4. Finished stock location
5. Pulled-from-stock location

The part number is self-explanatory. This is the same as the number of the part in inventory and the move card on the parts box which orders replacements. Each box of parts has a Kanban card with it.

The container capacity is the number of parts in the standard container. Each part has a fixed container quantity and a standard container. The quantity is small to simplify counting.

Each Kanban has a serial number to identify each card released. The number of cards released controls the quantity of product in inventory by limiting the number of containers that can be filled. Each container of product must have a card. The fewer the cards, the fewer the full containers. Multiply the card release quantity by the standard container quantity to get the maximum WIP stock.

The finished stock location data show where to put the parts; the pulled-from-stock location shows where to get them. For the work center, move cards transfer parts from parts storage to work-in-process storage at the work center.

Dispatch cards authorize manufacturing to produce products. Dispatch cards travel from storage locations to manufacturing work centers. Kanban dispatch cards contain the following data:

1. Part number
2. Work center number
3. Container capacity
4. Finished storage location
5. Raw materials required for production

The work center and part number provide routing information. This describes what is to be produced and which center will do it. The container capacity is the quantity of product to put in each container, with one card required for each container. The finished stock location indicates where the part is going when done. The raw materials and quantities are specified so move cards obtain the correct parts. Dispatch cards are both a single level bill of materials and a routing for a part. Figure 11-11 is a diagram of the flow

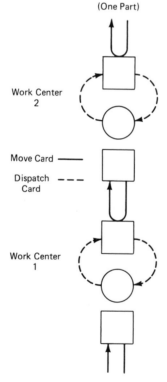

Figure 11-11. Single Part Kanban Flow

(Courtesy of George Plossl Educational Services, Inc.)

of Kanban cards between two work centers producing a single part. Figure 11-12 shows a multiple-part Kanban-controlled environment.

Kanban cards are the paperwork to move, prioritize, and control the flow of production. They contain the bill of materials and the routings. This eliminates computerized dispatch lists and prioritizes production based on consumed products. Kanban produces no forecasts for material requirements or other planning information since it is an execution, not a planning technique.

Kanban is a task master. To use this technique, the routings and bills of material must be 100 percent correct. Containers must be standardized to hold small lot quantities of parts. There must be a small number of Kanban cards released. Dispatching must match Kanban flow. Lines must balance, operations must overlap, and setups must be fast. Production must be reliable and

Figure 11-12. Multiple Part Kanban Flow

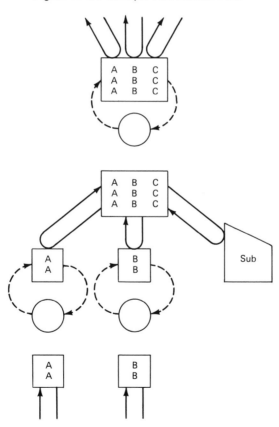

(Courtesy of George Plossl Educational Services, Inc.)

quality high. The master production schedule must be smooth and level, and capacity must be adequate and planned.

Kanban was developed by Toyota Motors in Japan. They admit that it took under 25 minutes to explain but took them over 25 years to implement. It will be no easier in the United States and Europe. The future is clear: Manufacturing control systems will provide material and capacity planning and control, while pull production will control execution. Engineering in a pull production environment will spend little time on standards but much time on the plant floor smoothing the flow, assisting workers to eliminate problems, and improving process quality and reliability.

12

PLANT ENGINEERING

PLANT LAYOUTS

Plant layouts are important to manufacturing engineering for studying equipment locations and material flow. Plant engineering needs layouts for power distribution, sewer, water, heating and air conditioning, fire protection sprinkler piping, and other utility locations. Both of these groups have an interest in keeping these vital records accurate and updated. Chapter 14, "Long-Range Facilities Planning," explains how plant layouts are crucial to linking business plans with space needs. In all of these areas, accurate plant layouts are crucial to manufacturing.

Because of the dual interest of plant and manufacturing engineering in plant layouts, there are disputes over layout ownership. These disputes result from differences in scale standards and assignment of responsibility for the layouts. Manufacturing engineering prefers a scale of ½ inch to the foot for machine locations. Plant engineering prefers a smaller scale—⅛ or ¹⁄₁₆ to the foot—so the whole plant is on a single drawing. All layouts should be made to some standard scale regardless of which department makes them. Have both departments agree on a standard scale and have anyone making a layout use it. A standard scale for all layouts permits sepia overlays of utilities over equipment, common drafting media, common storage, and easy duplication.

The most common standard scale is ¼ inch to the foot. This scale is accurate enough to locate machines on the layouts within three inches of their actual floor locations. It is accurate enough for plant engineering to locate shutoff valves, busway positions, and even approximate pipe and duct sizes.

Smaller scale layouts don't have this degree of accuracy and, while larger scale layouts would provide more accuracy, the layouts become excessively large.

Modern CAD/CAM systems provide the capability of changing scale for different users. The difficulty with CAD/CAM system layouts is that these systems are for design engineering use, and time is not available to put plant layouts on file. Plant layouts provide design engineering with insight into manufacturing operations, an opportunity design has not used to advantage in the past. If these layouts are put on design engineering CAD/CAM systems, the result is better communications between departments.

LAYOUT RESPONSIBILITY

Manufacturing engineering should have responsibility for plant equipment layouts. The manufacturing engineering department has the need for material flow and equipment location information and also the responsibility for layout data maintenance on equipment. Plant engineering needs utility locations and energy distribution information, so this group retains responsibility for utility layout data. Equipment and utility layouts are both keyed to column and wall locations. If the scale for both these layouts is common, the utility layouts can be on one drawing and the equipment on another, but the two can be overlayed for combined information. CAD/CAM systems make overlaying of utilities and machinery layouts easy. Employ CAD/CAM for layouts whenever it is practical.

MAINTENANCE

In the United States and Europe, plant engineering has traditionally had responsibility for the maintenance department, but manufacturing engineering has this responsibility in Japan. The difference is in the perception of the maintenance function. The Japanese believe that manufacturing equipment maintenance and preventive maintenance is the responsibility of the machine operator. They train operators to maintain their own machines and enlist specialists from manufacturing engineering only for assistance with electronics and sophisticated control systems. By contrast, Americans and Europeans believe that maintenance is a specialty performed by specially trained maintenance personnel. Since specially trained maintenance personnel work on buildings and other facilities, plant engineering runs equipment maintenance as well. Whether or not routine maintenance requires specially trained personnel is the fundamental difference between the two viewpoints. Both are valid but specialization doesn't promote functional integration. Companies using maintenance specialization must work harder to maintain good communications between plant and manufacturing engineering.

The manufacturing engineering department in Japan takes responsibility for buildings and facilities because it is their belief that the buildings exist solely to serve manufacturing. This belief produces spartan facilities designed around specific processes. The American view is that buildings are long-lived flexible assets; as such they should require little maintenance, they should reflect the character of the company, and their design should accommodate processes of many different kinds. These differing views account for the organizational differences and the diversity of the two approaches to facilities design.

The buildings and support facilities for American manufacturing are long-term depreciable assets. The U.S. government dictates that buildings depreciate in no fewer than 30 years. This accounting view and government policy encourages building designs that are durable and flexible. Construction is frequently concrete, brick, steel, and glass. Facilities are architectural design wonders. New facilities are designed with state-of-the-art solar heating and air conditioning systems, elegant designs encased in glass. Manufacturing buildings in the United States are as important as the processes they serve.

The spartan view of buildings is worth understanding. Many believe that process requirements and protection against the elements are all that matters. If it never rained or snowed and sunlight had no effect, no building would be necessary. Taken to extreme, this philosophy leads to tin sheds with portable kerosene heaters and roof ventilators for air conditioning. This type of facility will not win any architectural awards but serves the utilitarian function of protecting workers and manufacturing equipment from the elements. This type of building also limits the need for highly specialized maintenance staffs to keep complex environmental heating, ventilating, and air conditioning (HVAC) systems and controls operating. Both of these philosophies are found worldwide.

PRODUCTION MACHINERY

Production machine acquisitions are justified economically to increase the capacity to produce or to reduce costs. Capacity increases are justified by the use of the same or, better still, less labor. This emphasis on labor has produced some interesting results. Machines are acquired which produce items at a higher rate (pieces per hour) with fewer operators but have longer setup times. Setup time is less important as run length rises because the setup cost is spread over more parts. Running production equipment fast is paramount. Minimizing maintenance cost in the short term is the objective. This results in the philosophy, "If it ain't broke, don't fix it." Breakdowns are expected and manufacturing interruptions cushioned by inventory. Run speeds increase on subsequent machine purchases and run speed dominates thinking. It is important to have fast machines to catch up with production needs once the broken machine is fixed. This builds inventory, lengthens lead times, and causes large lot production.

There is a second philosophy. This philosophy is to keep the process running making small lots. This philosophy builds little inventory, provides smooth product flows, levels plant loads, and shortens lead times. It encourages new machines but keeps the old ones with permanent setups for odd-ball parts production. In short, it improves productivity. It is this philosophy that has produced the greatest improvements in inventory turns and profits. It is this philosophy, however, that has the greatest effect on maintenance activities in terms of effective preventive maintenance rather than rapid breakdown maintenance.

BREAKDOWN MAINTENANCE

The greatest cost of operating a breakdown maintenance system is the cost of lost capacity. The equipment is not available to produce when the plant requires production. Inventory covers these disruptions but the costs in lead time, customer service, and inventory investment are high. When major setup reduction projects are started to reduce run quantities and WIP inventories, breakdowns become a major problem. Whole plants shut down when capacity is interrupted. Breakdown consequences are visible immediately in a process operation such as a refinery or chemical plant. Breakdowns in a process plant shut down the whole plant and therefore must be avoided if at all possible.

Breakdown maintenance fails in the process environment because it doesn't emphasize keeping the plant running. Preventive maintenance works to eliminate breakdowns. Batch manufacturing tolerates breakdown maintenance by increasing inventory to compensate. The price for a breakdown maintenance philosophy is high inventory, long lead times, and lower customer service levels.

PREVENTIVE MAINTENANCE

The majority of breakdowns are preventable and, of the remainder, many are predictable. Only a small percentage of breakdowns just happen. These facts are supported by surveys of manufacturing failure causes. Figure 12-1, which details the causes of failure in heavy machinery, is the result of a survey of several hundred manufacturers using heavy machinery.

Machine overloads are frequently the direct result of improper adjustment during setup. Standardizing these adjustments can make this failure rate drop significantly. Design changes can incorporate shear pins or other over-

Figure 12-1. Heavy Machinery Failure Causes

Machine Overload	28%
Faulty Die or Product Design	26%
Repair Shop Malpractice	17%
Improper grinding	
Stresses in welds	
Defective welds	
Faulty lubrication	
Misalignment	
Faulty Heat Treatment	17%
Stress Risers	11%
Sharp corners	
Wrong fillets	
Wrong Material	5%

load protection devices which minimize the effects of overloads that do occur. Thus, a high percentage of these failures are preventable.

Faulty die and product designs are the cause of a significant percentage of heavy machinery failures. Failures directly attributable to dies or product designs that resulted in transfer tooling malfunctions were classified in this category. Dies that misalign during operation or jam cause overloads. Many of these problems can be eliminated by designing products for the processes and by standardizing die designs. Product designers can improve their designs to prevent equipment failures.

Shop malpractice can be eliminated by adequate training and supervision. If the maintenance staff has no ongoing training program to improve and practice their skills, the results will be demonstrated here. Maintenance must be supplied with the correct tools to do the kind of repairs that prevent future failures.

Faulty heat treatment of shafts and other machine parts is not preventable in purchased machines; however, it is preventable when the parts are repaired or replaced. It is suggested that all heat-treated repair parts be inspected prior to replacement.

Stress risers are preventable also. Failure inspections by engineering to determine the nature of the failure and the remedy should be automatic. Sharp corners and fillets should be eliminated to prevent failures in machine parts even if those parts are manufacturer supplied. There is little reason for continuing to accept machine failures just because the equipment manufacturer may not have designed for failure prevention. Nothing is sacred about machine design except that if it can be improved, it should be.

The wrong material is usually the manufacturer's problem, and these failures constitute only 5 percent of the total. If 95 percent of the failures in heavy machinery are addressed, inform the manufacturer of these defects. Manufacturers rarely want failures they can eliminate.

GEAR FAILURES

Among the major components of machines are gears. Figure 12-2 shows the results of gear failure studies.

Gear failures are not inevitable. Surface problems can be detected before actual failure occurs. Periodic inspection of gears can record the very beginnings of these problems, and replacements can be installed before failure occurs. These problems are not preventable but downtime because of them is. If failures are predicted, replacements can occur before failures cause capacity loss.

Bending fatigue is not preventable but historical data provide an estimated life so that failures may be predicted. As with surface problems, these failures need not result in downtime.

Impact loads are the result of machine jams and overloads which are preventable. Cyclical impact loads from machine cycles are predictable, so these failures need not cause machine downtime either.

Even if all 23 percent of the miscellaneous failures are neither predictable nor preventable, the other 77 percent are addressable. This means that three out of four gear failures need not cause production downtime.

Figure 12-2. Gear Failure Causes

Surface Problems	38%
20% pitting, spalling, crushing	
18% scoring, scuffing	
Bending Fatigue	24%
Impact Loading	15%
Miscellaneous	23%

ROLLING ELEMENT BEARING FAILURES

Bearings are the heart of all rotating and reciprocating machines. Bearing failures represent over half the machine downtime in most manufacturing plants. A majority of this downtime is preventable. Figure 12-3 shows a summary of a survey of rolling element bearing failure causes.

Almost half the failures are caused by lubricant problems. These are either preventable or predictable. Adequate supplies of lubricant can be assured if lubrication is part of the design, as when a higher volume pump supplies more lubricant. Of course, bearing failures must be properly identified to be prevented. If the cause is attributed to the wrong lubricant rather than inadequate supply, the bearing failure may change but the life may remain the same.

Many oil companies provide a free analysis service to businesses using their products. Oil companies will make a survey of the equipment and the lubricants in use. They will also analyze failures for lubricant causes and analyze samples from machines after a failure. They will then make recommen-

Figure 12-3. Rolling Element Bearing Failure Causes

Lubricant Failure	47%
Inadequate supply	
Wrong lubricant	
Lubricant breakdown	
Foreign Material	18%
Dirt	
Metal particles	
Brinelling	12%
Faulty assembly	
Shock loads	
Stress Concentration	10%
Misalignment	
Excessive shaft deflection	
All Others	13%

dations on the correct lubricants and on frequency of lubricant changes to maximize machine uptime. There is no excuse for continued lubricant problems except for inadequate attention, lack of preventive maintenance, and improper training.

Foreign material in bearings is a problem in dusty atmospheres such as around rock crushers and powdered chemical handling applications. Bearing companies are ready to assist manufacturers with improved seal and shield designs, lubricant recommendations, and design advice to extend bearing life in such applications. These companies also provide bearing failure analysis to determine what to do to either extend life or predict failures. If the failure can be predicted, the downtime can be scheduled when most convenient.

Faulty assembly and shock loads are a function of either overloads or inadequately trained maintenance personnel. Both of these problems are preventable, as are stress concentrations. Excessive deflections are due to overloads and misalignments are due primarily to poor maintenance. Again, these failures are preventable.

EQUIPMENT DOWNTIME

Over 90 percent of the heavy machinery failures, 75 percent of the gear failures, and 85 percent of the bearing failures need not result in production downtime. This means that over 80 percent of the downtime in manufacturing today can be eliminated. To accomplish this, a comprehensive preventive maintenance program must be started. This program must contain four elements:

1. Identification of failure causes
2. Education of maintenance personnel
3. Elimination of preventable failures
4. Prediction of nonpreventable failures

THE PREVENTIVE MAINTENANCE PROGRAM

The first step in developing an effective program is to start a failure classification system. Every failure is predictable, preventable, or neither. The neither category is rather small and expert help may be required from equipment manufacturers and electronics companies to shift failures in this category into one of the other two. The first task after classification is to educate maintenance personnel to recognize the causes of failures and to formulate methods to prevent or predict the others.

Elimination of preventable failures takes action. Employee disciplinary action for not stopping preventable failures will not cure the problem. Machine operators and maintenance must work together to prevent downtime. Operators must be alert and encouraged to detect and report incipient trouble signals—heat, noise, vibration, and so forth. Preventive maintenance must be scheduled so that manufacturing equipment is available to be repaired when maintenance people can be scheduled to work on it. The manufacturing control system can schedule work orders for maintenance as well as production. This must be part of any preventive maintenance program.

Failure prediction is as much a part of a preventive maintenance program as is failure prevention. Relays have a rated life in cycles. Once a relay cycles that many times, failures occur at an ever increasing rate. Only about 10 percent of the relays will fail prior to that number of cycles because the rating is computed statistically. If the number of cycles are monitored, the relay can be replaced at the 10 percent failure level, and 90 percent of the relay failures disappear. Bearings are rated the same way. There are special sonic testers that detect incipient bearing failure so that they can be replaced before actual failure occurs. These predictive solutions to downtime make use of failure forecasting techniques not unlike the forecasting techniques used to predict sales volumes.

FAILURE PREDICTION

There are primarily three bases for predicting failures: historical performance, design life estimates, and inspection. Bearing, gear, and steam trap failures can be forecast by inspection. By inspecting these parts on schedule, the signs of beginning failure are spotted and repairs scheduled before failure occurs. This inspection failure prediction system is not applicable to many parts and requires monitoring on a regular basis to work. Although effective, the inspection system is primarily used in conjunction with instruments as test or inspection devices. The most notable of these is the infrared steam trap thermometer. As the trap fails, it will begin to heat up. The thermometer will detect the higher temperature and the trap can be scheduled for repair.

Another example of inspection testing is the bearing noise failure detection system. SKF developed a correlation between bearing noise and failure. By monitoring both the frequency and amplitude of the noise developed by rolling element bearings, it is possible to detect failure before it occurs. There is a tester on the market today which measures the noise produced by running bearings and predicts failure. This inspection system is effective for some applications and yields the best hope of getting the closest to the maximum part life.

The second basis for predicting failure is the design life estimate. The most notable application of this is in bearings. Rolling element bearings are tested under load, and life data are accumulated. It is then possible to use the load and life data to compute a life, called the B-10 life, of a whole population of bearings. This B-10 life represents the life in hours at which 10 percent of all the bearings will fail under any given load. If the load can be measured or computed, the B-10 life can predict failure. This is a design life basis because the design load or an estimate of the load is required to compute the bearing life in use. This basis can also be applied to any item given a design life.

The most frequently employed system for failure prediction is historical performance analysis. This system works on the assumption that failures will result at regular intervals. Past history is the basis for the failure prediction. The system usually employed is to calculate the mean-time-between-failures or MTBF for the part. Given sufficient history, that mean is fairly constant, and replacement or repairs are scheduled prior to reaching that mean time. This is an effective system but history must be accumulated on the parts.

This MTBF system predicts the schedule for preventive maintenance on large, complex machines containing hundreds of individual parts. Here the whole machine failure rate is determined and a thorough maintenance procedure is scheduled for that machine before failure has been predicted to occur. For some machines, this prevention maintenance program is published by the machine manufacturer. The best example of this is the automobile. Every so many miles certain maintenance is recommended. This keeps the car running at peak performance and prevents unexpected failure. For many production machines, programs of this kind are not available from the manufacturer. In these cases, the failures of machines and the causes of those failures must be tracked historically, and plant engineering needs to develop an effective schedule.

Most equipment manufacturers do supply some information on preventive maintenance programs even if this is limited to lubrication frequencies. By watching the types and frequencies of failures, engineering can develop a more comprehensive program for failure avoidance. For example, suppose a machine fails about every 2000 hours due to drive V-belt failure. Let's assume that the machine operates on eight-hour shifts when running. Then we can establish a schedule of running 1920 hours—or every 240 shifts—to replace the V-belts. After this, the frequency of belt failure should be much lower and the machine more reliable. A thorough discussion of forecasting techniques used to predict failures is found in Chapter 13. The fundamental goal of any preventive maintenance program is to reduce the failure rate and increase the uptime available to production.

BREAKDOWN COST

Breakdowns are costly. Unless significant excess capacity exists in a process, inventory will be high or jobs will be late. If enough capacity exists in a work center, overtime and multiple-shift operations can be used to catch up. Both of these alternatives are costly in terms of inventory or excess capital in equipment. Breakdowns also have a lengthening effect on lead time, which increases costs. As breakdowns delay production, lead times extend to insure delivery on schedule. Chapter 10, "Lead Time and Setup Time," covered the results of lengthening lead times and the effects on manufacturing. Breakdowns are inevitable in many batch manufacturing plants, while they are preventable in process plants. If batch manufacturing is to operate in a "just-in-time" or semi-process environment, breakdowns must be prevented.

ENERGY AND ENERGY CONTROLS

Plant engineering conducts energy conservation programs. This is because insulation of buildings and energy distribution are under their control. In order for engineering to track consumption and to control energy effectively, accounting must help by providing copies of all energy bills. Although energy control and energy cost control are not synonymous, the first place to start a conservation project is with an understanding of the billing structure.

Energy Billing Structures

Many businesses assume that electricity is billed at a rate per hour measured in kilowatts. Most commercial rate structures are not billed that way exclusively. A standard method for commercial account billing has some definitions worth noting.

1. Contract Capacity: This is the amount of energy in kilowatts the company can draw if they drew power at the maximum rate. This contract capacity is the maximum number of kilowatts the power company will supply at peak demand.
2. Peak Demand Kilowatts: This is the number of kilowatts demanded dur-

ing some predefined time period. This is a measure of the peak plant usage of electrical energy.

3. Demand Billing Period: This is the time period for which the average of peak demand kilowatts is measured. This period can be as long as an hour or as short as a few minutes. The normal periods are 15 minutes or 30 minutes.

4. Time-of-Day Usage Rate: The power company may provide energy at a different cost per kilowatt hour based upon the time of day in which the power is used.

5. Usage Rate: This is the cost per kilowatt hour charged for the energy consumed.

6. Power Factor Penalty: This is a surcharge imposed by the power company for having to supply power factor correction. This is usually a percentage of the energy charge based upon a power factor range and a measurement of the average or peak factor.

Contract Capacity

Power companies must supply power to all customers during the period of peak demand. Their power generating plants and distribution system must accommodate these peak loads. Thus, the rate at which a user can demand power has a direct correlation with power company costs. If the power company does not adjust capacity to meet load, customers face voltage reductions, called brownouts, or outright system collapse, called blackouts. To anticipate the demand in time to acquire capacity in generating plants and distribution systems, the power company contracts with its larger customers for capacity. This is the purpose for the contract capacity. The power company wants this capacity value as high as possible for demand billing as well as forecasting.

The utility would like the contract capacity to be the sum of the connected loads for everything in the plant. The connected load for a device is the maximum current the device will draw. For motors and other items with high starting currents, this is far above running power consumption. For this reason, companies negotiate with the power company to arrive at contract capacity. Although the utility wants connected load, the company can frequently contract for less than the total connected load. Since this contract capacity is used to compute a minimum demand billing, it is in the manufacturing plant's best interest to keep that capacity just slightly above its instantaneous peak demand. The power company may resist any changes in this contract capacity, but as a plant grows, its usage will begin approaching the contract capacity and new contracts must be negotiated to protect the power company network and to avoid penalizing the large user unnecessarily.

PEAK DEMAND KILOWATTS

Power utilities bill based upon the power demanded by large customers. This is due to the amount of generating capacity the utility must bring on-line to support that power level. The power demand bill is usually a flat rate per kilowatt, and the demand level is determined by the average rate over the demand period. For example, assume that a rate of $6.00 per kilowatt demanded is charged for a demand period of 15 minutes. If the demand in kilowatts fluctuates between 1150 and 1250 during any 15-minute period during a month and the demand average during that 15-minute period is 1200, the demand billing will be 1200 × $6.00 or $7200 for that month. If that power level is below some minimum percentage of the contract capacity, usually 65 percent, the minimum demand billing will be for the minimum. Thus, if the contract capacity in this case was 1900 kilowatts, the minimum billing would be .65 × 1900 × $6.00, which is $7410.

The path to lower demand energy bills should now be clear; reduce demand levels, demand constant levels, and keep the contract capacity down. The contract capacity has little influence on the bill if the minimum draw is always above the minimum percentage of contract used for billing. Thus, the goal should be to keep the capacity high enough to provide adequate business growth and still maintain actual demand above that minimum percentage of the contract.

Time-of-Day Billing

The priority for power companies is to supply all their customers with adequate power. The peak power plant loads frequently occur between certain hours of the day. In order for the power company to provide an incentive for users to move power consumption from periods of high usage to periods of lower usage, a time-of-day rate has been established in many places. This rate is reflected in different per kilowatt hour charges for power. Reducing usage during peak hours and shifting that usage to off-peak hours results in energy cost savings without actually reducing consumption.

Power Factor Penalties

The power factor is technically the cosine of the phase angle between voltage and current in a power line. Electric motors use power to produce a magnetic field and this causes an inductive shift in the phase angle. For businesses with many motors, the cumulative shift can be significant. To offset that shift, capacitors are put on the power line. If these capacitors are not put on by the large user, the power company must correct the problem and they charge a premium to do so. The best policy is to put power factor correction capacitors on every motor above 100 horsepower. Connect these capacitors downstream of the contactor so they come on line with the motors. General correction is applied to keep the power factor above .90 to maximize main power transformer capacity and minimize penalty billing.

Other Energy Sources

Oil, coal, propane, diesel fuel, and natural gas are all used as energy supplies. Billing for these items as well as for electrical power may show ways to reduce the cost without reducing consumption. Again, the first goal of an energy program should be to minimize cost for the energy used. This provides a basis for proposing energy reductions that save energy and lower costs to the consumer.

THE ENERGY AUDIT

The energy audit is the same as a financial audit. The goal is to determine the energy balance of total inputs by type and the sources of consumption and losses. There are two usage categories for energy—environmental and process. Environmental uses are space heating, ventilating, air conditioning, and lighting. Process uses are drive motors, process heating and cooling, material handling, and process liquid circulation pumping. The energy audit should define these separately. Process energy usage is required for manufacturing to continue, but losses of energy by processes can be reduced. The greatest potential for conservation is frequently environmental.

The energy audit captures every source and use of energy. This means that each process or machine should have its energy consumption determined. The total plant energy input equals the total energy lost, or the plant temperature will rise until a balance is reached. During cold weather the process energy generates heat which warms the environment, reducing the energy consumption for heating. During warm weather, the process adds heat which must be removed by ventilation or air conditioning for temperature control. A process in which thermal energy is lost helps in winter but increases air conditioning requirements during the summer. The energy audit serves as the basis for an energy conservation and cost reduction program and as the basis for an energy resource planning program.

THE ENERGY RESOURCE PLAN

Energy use is the result of production and environmental consumption. Heating, air conditioning, and ventilation use energy for environmental control, and the manufacturing process consumes energy to produce the product. Process energy consumption is the least controllable. Conservation projects reduce process consumption, but shutting down process consumption stops product production. There are few companies that tolerate such drastic energy reduction programs. More commonly, peak loads shift in time by moving some processes to off-peak hours. Computer-controlled time shifting can be done within a time period by scheduling usage for noncritical energy consumers. These noncritical users are glue pots, heat treatment bath heaters, plating tank heaters, and similar usages. These applications can be scheduled to retain temperature without having all heaters on at the same time. Energy reduction programs for production processes are limited to energy efficiency improvements and noncritical user scheduling.

Process energy needs tie in to manufacturing so that scheduling can be projected to include energy usage. Process energy ties directly to the production plan of the business and can be estimated using a resource planning system. Process energy requirements are planned from the product routings and process energy consumption of processes. These data provide the information to compute expected energy usage by work center by time period (usually weekly) and also give forward visibility into energy requirements as expressed by rough-cut energy requirements planning (ERP). Rough-cut ERP provides forward visibility into needs for new substations, higher capacity busways, larger gas distribution piping, and the like.

Process ERP also permits environmental energy planning. ERP permits the energy plant balance to be computed. During the heating season, processes consume energy and release it to the environment. This is process energy in. The outside temperature, wind, and other factors determine the rate of energy loss—energy out. The balance is maintained by adding environmental heating. Air conditioning and summer ventilation work in reverse to remove process energy and heat input from the outside. Again, a balance can be written.

Modern long-range weather forecasts permit forecasts of the energy requirements needed to support the environment and the processes. This is a better way to budget energy than to use past history with a fudge factor. It also focuses attention on the rate of heat loss or gain through the building envelope. Better insulation programs depend on reductions of energy through the envelope for justification. ERP, for both process and environment, provides the data for justification and monitoring of success.

Plant engineering needs ERP data and supplies energy usage facts to the business computer system to calculate the results. The time has past when energy was cheap. It is a resource like any other in the manufacturing plant. It is controllable. Management needs total resource planning, and plant engineering holds the keys to effective energy management. Of course, the engineering group must have support from the business computer because the master production schedule provides the data for process ERP. Plant engineering must participate by helping to run the business with planning for energy and maintenance.

13

FORECASTING
AND MACHINE UPTIME

Manufacturing planning and control are future oriented. The past is beyond control, the future must be prepared for. The heart of this preparation is an estimate, guess, or assumption about future conditions. Forecasting is the name given the process of preparing this estimate of the future. Figure 13-1 shows the general types of forecasts prepared to run a manufacturing business.

These types break down into four forecasting time periods. Long-range forecasts cover the period from five years and beyond. Intermediate-range forecasts cover the six-month to four-year horizon. Short-range forecasts typically cover the period from one month to six months. Immediate-future forecasts cover the one-week to one-month horizon. The classifications of short versus long depend on industry. A defense contractor making airplanes sees short range as the next two years or the duration of the contract. For a computer manufacturer with product life cycles of 18 months, long range may be one year or less. The above classification of range is typical for most consumer products manufacturers.

From these typical business forecasts, many subsidiary plans develop: inventory balances, cash flow, repair parts, energy requirements, square footage in facilities, and equipment life, to name a few. These subsidiary forecasts serve specific users' needs and must be integrated with the business forecasts for users to gain an overall view of the business plans. Perhaps the most important principle of forecasting was stated by George Plossl: "A forecast is a set of numbers to work from, not to." This states most clearly that forecasts are not goals. Forecasts are estimates of what is expected to happen in the

Figure 13-1. General Types of Business Forecasts

Forecast	Used by
Population Growth	Marketing to determine market growth
Production Growth (5-year plan)	Manufacturing and engineering to plan for process and facilities expansion
Production Requirements Hours (2-year plan)	Manufacturing and finance to establish gross yearly budgets
Sales by Product Family (Next 12 months)	Sales to set quotas
	Finance to set expense budgets
	Manufacturing to compute machine and worker output capacity
	Material control to compute gross raw material needs and inventory levels
	Engineering to compute energy requirements, gross repair parts needs, preventive maintenance levels by work center, etc.
Sales Next Quarter	Material control to compute work center and vendor required capacities
	Engineering to compute energy requirements, gross repair parts needs, preventive maintenance levels by work center, and so on
Sales Next Week (by item)	Material control to set assembly schedules and dispatching priorities
	Engineering and material control to schedule PM and prototype manufacturing needs

future and plans are developed to respond to these predictions. As forecasts change, so must the plans.

FORECAST CHARACTERISTICS

There are four fundamental characteristics of all forecasts:

1. Forecasts will be wrong.
2. Forecasts are more useful with error estimates.
3. Forecasts are more accurate for aggregates.
4. Forecasts are more accurate over short horizons.

Forecasts, being predictions of the future, always contain errors. Regardless of how hard one tries to predict accurately by means of complex mathematical techniques, errors will still be present. The law of diminishing returns applies to obtaining accuracy. The more accurate the forecast must be, the more it costs, exponentially. For example, the cost of estimating project costs is a function of the accuracy required. A ± 30 percent accuracy range can be achieved for 0.5 to 1 percent of the total project cost. A ±10 percent accuracy range will cost 1 to 2 percent of the project total. A ±5 percent range will cost 4 to 5 percent of the total project cost. It costs as much as the project to obtain an accuracy range of under 0.25 percent. This is not cost effective. It is far more profitable to develop flexibility to cope with errors. The key is to develop a forecast which includes a measure of its inaccuracy.

Forecasts having error estimates provide the basis for contingency planning. Using upper and lower bounds for forecasts permits users to plan for these different forecast levels. Bounds can be determined from observations of history, by an educated guess, or by some combination of the two. Statistical methods are applied to determine accuracy as well. The fact remains that forecasts are more useful for contingency planning if their accuracy is stated over a range.

Aggregates provide greater forecast accuracy. In 1923, the engineers in Western Electric developed the first set of quality inspection sampling tables. These tables provided estimates of the number of defects in a lot based upon a sample lot of items checked. This approach is still a mainstay of statistical quality control. It is difficult to determine which item in a lot has a defect, but the total number of defects in the lot can be determined accurately by sampling. This technique is also useful for forecasts. The forecast of total sales volume of all products is always better than the estimate for any one item. These single-item variations can be dramatic. For one company, the total annual sales volume forecast accuracy was ±5 percent, while the typical error for individual items was ±800 percent!

The shorter the horizon, the better the forecast is. Five-day weather forecasts are detailed but monthly ones are only useful for general weather patterns. Sales forecasts are better for next week's sales than for next month's. In general, forecast accuracy is a function of horizon. The longer the horizon, the higher the probability of error and the greater the error magnitude.

STATISTICAL FORECASTS

Statistical techniques treat the basic elements (trend, seasonality, and randomness) separately. These three components are shown in Figure 13-2 for the growth of sales in an expanding manufacturing company.

The trend is upward and the formula for the best straight line is computed by the least-square error technique. The slope can be computed by the formula:

$$\text{slope} = \frac{n\Sigma(xy) - \Sigma x \, \Sigma y}{n\Sigma X^2 - (\Sigma x)^2}$$

The Y intercept is computed by the formula:

Figure 13-2. Manufacturing Company Sales Growth

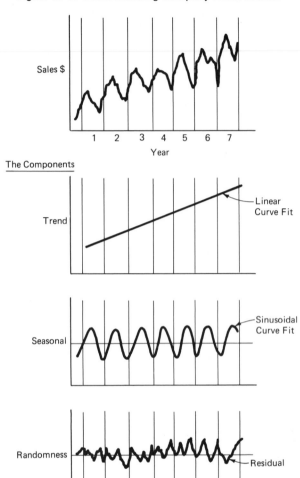

$$\text{Y intercept} = \frac{\Sigma y \Sigma x^2 - \Sigma x \Sigma xy}{n \Sigma x^2 - (\Sigma x)^2}$$

Modern personal computers with spreadsheet programs make this task simple. The great difficulty is to get current x and y data from the business computer to apply it effectively.

Once the trend is established, subtract the trend y value at each point from the actual value to remove the trend. This leaves the seasonal pattern and the scatter residual in the data. Seasonality is often assumed to be sinusoidal, and the residual is determined by subtracting the seasonal value computed from a formula of the form:

$$y = A \sin Bx$$

where

A is the amplitude of the seasonal variation
B is the wavelength or period of the seasonality

THE AVERAGE AND WEIGHTED AVERAGE TECHNIQUES

Two statistics are commonly used in forecasting—the running average and the weighted average. The running average is calculated using the formula:

$$\text{Forecast} = \frac{\Sigma y - y_1}{n}$$

where

y is the value being forecast
n is the number of periods

This technique is useless for forecasts involving trends or seasonality because it fails to account for these variables. However, it is useful for forecasting such items as gasoline mileage per gallon (MPG), mean time between failures (MTBF) of machinery, process energy usage rates per hour, and so forth. It can also be used on data once trends and seasonality have been removed. The major drawback to this technique is the volume of data required to use it. A 12-month running average needs 12 monthly numbers stored for each item forecast. For many items, this requires a large amount of data.

THE WEIGHTED AVERAGE (EXPONENTIAL SMOOTHING)

The simple running average gives equal weight to new data and old. A 12-month running average weighs the last month equally with all the others—that is, one-twelfth of the year. For some applications, this produces a forecast which is too sensitive to current data. Suppose engineering needs a forecast for mean time between failures (MTBF) on a process. Assume the old average value was 236 hours but breakdowns are erratic. A simple average of the old value with the new data is too erratic to be useful.

```
Simple Average
Old Forecast           = 236
Actual this Occurrence = 155
                         391
Simple Average  391/2 = 195.5
```

The simple average of approximately 196 hours is too low. The real value is closer to 236 hours.

Weighted Average

		Weight	Product
Old Forecast	= 236	.9	212.4
Actual this Occurrence	= 155	.1	15.5
		1.00	227.9

This weighting factor technique supresses the sensitivity to new data and provides a more useful forecast. In general, the formula for the weighted average is:

$$\text{New Forecast} = F(\text{actual}) + (1 - F)(\text{old forecast})$$

where

F = weight factor

This technique approximates the results of a running average by setting the weight factor appropriately. For a 12-month running average approximation, F is $2/(12 + 1)$ or approximately 0.15. The formula for computing the F factor for running average conversion is:

$$F = \frac{2}{n + 1}$$

This reduces the need to store the data to use the running average and is far more practical where large volumes of data are required.

This weighted average technique is called first-order exponential smoothing. The formula is rearranged and presented as follows:

$$\text{New Forecast} = \text{Old Forecast} + F(\text{actual} - \text{old forecast})$$

The difficulty with using running averages and weighted averages is the inability to get a good forecast from sparse data. This is the case when engineering starts a program to measure MTBF on equipment. The number of data elements is few and the exponential smoothing and running average techniques are too slow to use to approximate the MTBF quickly. R. E. Kalman developed a new optimal recursive estimation method in 1960. His new technique bears his name, Kalman filtering.

SECOND-ORDER EXPONENTIAL SMOOTHING

Second-order exponential smoothing is designed for data having a trend. The first-order technique, or the Kalman filter, computes the first factor based upon the following equation:

$$\text{Factor A} = \text{old forecast} + F (\text{actual} - \text{old forecast})$$

where

F is the weight, or Kalman gain factor

The second factor estimates the lag and corrects the forecast for the trend. The factor B uses the same equation as first-order smoothing but different values. The equation is:

$$\text{Factor B} = \text{old B forecast} + F (\text{Factor A} - \text{old B forecast})$$

Note that Factor A is treated as the actual in this second equation. The value of F is usually the same for both factors, but Kalman filtering on both is rational where trends vary. The new forecast is now given by:

$$\text{New Forecast} = 2 \text{ Factor A} - \text{Factor B}$$

This technique is inaccurate when the forecast is very low or when it is zero in certain periods. Seasonality and data without trends also introduce forecast inaccuracies.

KALMAN FILTERING ESTIMATION

Multiple-variable Kalman filtering problems are routinely solved on today's computers. The bibliographic references contain a complete discussion of the technique. The discussion here is limited to single-variable linear estimates, those that are candidates for exponential smoothing techniques or running averages. The example used is the computation of mean time between failures (MTBF) for a factory process.

The Kalman technique uses a small sample of actual readings to predict the value of a variable in a noisy environment. It reuses the past forecast to predict the future and is thus recursive. For MTBF, the assumption is that a mean value exists for failure for a process from all causes. The second assumption is that the failure distribution follows the normal distribution function. The mean is defined as:

$$\text{MTBF} = \frac{\Sigma \text{ (times between failures)}}{\text{number of occurrences}}$$

This property is difficult to measure without many actual measurements. In the MTBF case, the number of data points is limited by the number of actual failures documented. The greater the MTBF, the sparser the data become. The MTBF is a constant and the Kalman technique needs to estimate that value. If MTBF is constant, the sample data will have a variance (q) around the mean and the following equation applies:

$$\text{FTBF}_1 = \text{FTBF}_0 + W$$

where

> $FTBF_0$ = Next Forecast Time Between Failures
> $FTBF_0$ = Last Forecast TBF
> W = a random sample from a zero-mean noise universe with variance q

This means that forecasts will vary because of some random noise. This noise comes from random disturbances within the process and is really process noise. The measured values of TBF are related to the forecast by:

$$ATBF = FTBF_0 + V$$

where

> $ATBF$ = Actual TBF measured this time
> $FTBF_0$ = Last Forecast TBF
> V = Measurement Noise having variance r

The forecast TBF is equal to the actual TBF within the limits of normal measurement scatter. The Kalman technique estimates the TBF using the following formula:

$$FTBF_1 = FTBF_0 + K(ATBF - FTBF_0)$$

where

> $FTBF_1$ = New Forecast TBF
> $FTBF_0$ = Last Forecast TBF
> $ATBF$ = Actual TBF measured this time
> K = Kalman Gain Factor

This formula has a form exactly like the one for single-exponential smoothing except the factor K is dynamic. The K factor is a measure of the confidence placed in current data as it is for exponential smoothing. The advantage for a dynamic K value is that K adjusts as experience improves. The value of K is computed from:

$$K = \frac{P_0}{P_0 + r}$$

where

> P_0 = Last Error Covariance
> r = Variance of measurement noise

The error covariance P is an estimate of the variance in the Kalman filter's estimate of FTBF. The value of P provides an automatic estimate of the accuracy of the forecast. This error estimate is used to compute a new Kalman gain factor K by the following formula:

$$\text{New } P_0 = (1 - K) \text{ Old } P_0 + q$$

This makes the Kalman filter dynamic and recursive. Old data are used to make a new estimate with the gain being a function of the stability of the data. The value of r, measurement noise variance, controls the value of the Kalman gain factor K. If r is very large, failures have wide variance and the gain is low, placing little confidence in new values. If r is small, failure variance is small, and the gain places a lot of confidence in new data. In the limit, $r = 0$, the gain becomes 1 and the measurements are error free. The value of q prevents the filter from having a gain equal to 1 and completely ignoring the new data. The larger the value of q, the faster the filter responds to new data.

For the MTBF case, set q at 0.40 and r at .02 to start. Pick a starting value of P at 25 percent of the estimated MTBF. Seed the estimated MTBF with some rational value and the filter will approach the real MTBF within four failure time entries. Reduce the value of q to .01 for succeeding calculations to make the filter less sensitive.

WEIBULL FUNCTION FAILURE PREDICTION

This statistical technique was described by W. Weibull in Sweden in 1949. It is a technique used to sample a small population and describe the characteristics of a larger one. It assumes failure patterns in the sample population represent those of the entire population. Since this function is useful for determining the probability of failure of a population of similar items based on observations of a small group, it has been used in many failure testing programs.

The Weibull cumulative distribution function is:

$$\text{Weibull C.D.F.} = 1 - e^{-(x/s)^b}$$

where

$e =$ base of natural (Naperian) logarithms (2.718...)
$x =$ elapsed time
$s =$ characteristic life (the value of life at C.D.F. $= 63.2\%$)
$b =$ slope of the Weibull curve which describes the type of failure distribution

b Value	Type of Distribution
3.5	Normal
1.0	Exponential

The Weibull distribution requires no previous knowledge of the type of distribution because plots made on Weibull graph paper invariably plot as straight lines. The derivation of the formula can be found by referring to W. Weibull in the bibliography.

To use the Weibull function for estimation, data must be obtained on the elapsed time from test start until failure for a sample of similar items. The

Figure 13-3. Bearing Life Test Sample

Item Number	Cycles to Failure	Median Rank	5% Rank	95% Rank
1	326,800	0.056	0.004	0.221
2	445,600	0.137	0.031	0.339
3	554,400	0.218	0.072	0.438
4	610,700	0.298	0.125	0.527
5	739,900	0.379	0.182	0.609
6	814,700	0.460	0.247	0.685
7	929,000	0.540	0.315	0.754
8	1,006,900	0.621	0.391	0.818
9	1,222,100	0.701	0.473	0.876
10	Suspended at 1,222,100			
11	Suspended at 1,222,100			
12	Suspended at 1,222,100			

samples are tested to failure, the life is computed in hours, cycles, or similar units, and a median rank is determined for each sample. Median ranks for various sample sizes are available in table form. Median ranks are the cumulative percent failed values used for plotting the vertical axis of the Weibull plot. The horizontal Weibull axis is the life in cycles, hours, and so forth.

The Weibull plot computes the B10 life, or the life at 90 percent reliability. The forecast errors are plotted as the 95 percent and 5 percent confidence limits using the 95 percent and 5 percent ranks for the samples. Figure 13-3 shows a sample experiment of a bearing life test.

From the Weibull plot of these data, the B10 life is 370,000 cycles for this bearing under these test load conditions. The confidence limits are 95 percent = 160,000 cycles and 5 percent = 600,000 cycles. The Weibull estimation technique computed the number of cycles required to fail 10 percent of the bearings and the error range for this estimate.

This technique is useful for determining failure rates on machinery. Life calculations permit scheduled preventive maintenance on parts which regularly fail such as printed circuit boards, relays, and other similar parts. Replacement, rebuilding, redesign, or inspection reduces unexpected failures.

USING FORECASTS

The first axiom of forecasting is, ''Forecasts will be wrong.'' Highly precise calculations from complex computer programs follow Gallosis' revelation: ''If you put tomfoolery in a computer nothing comes out but tomfoolery. But this tomfoolery, having passed through a very expensive machine, is somehow ennobled and none dare criticize it.'' The first principle of forecasting is to limit forecast results to a maximum of three significant digits. For those who remember them, this used to be called slide rule accuracy. Do the calculations to double precision computer accuracy if forecasting mathematics require it, but present results using only the first three left digits of the result.

The second principle of forecasting is to use the simplest technique possible. There are some very complex mathematical forecasting models available in addition to the ones presented here. There are benefits and costs associated with all forecasting techniques. Choose the simplest technique that yields a usable forecast. The more complex the method, the longer it takes to run, but the results are still estimates subject to many errors.

The third principle of forecasting is to track actuals and compare them often with the forecasts. Revise the forecasts when they are no longer valid, but remember not to forget the fourth principle. It states, *Do not revise published forecasts until the new forecast will exceed the error tolerances of the old one.* There is no point in revising a forecast as an end in itself; thus it is important to establish a forecast tolerance using statistical significance criteria. If the new forecast has a statistical probability of over 50 percent that it is the same as the old one, don't publish the new one.

Judgment must be exercised as well. Mathematics won't predict an accurate MTBF when a machine has been redesigned to eliminate failures. Revise forecasts when significant changes are detected or are known to exist.

CHOOSING A TECHNIQUE

Choice of technique requires an understanding of the variable being forecast. For engineering failure and mean time between failure prediction, the variable has a near zero slope. These horizontal variables are best forecast with a running average, first-order exponential smoothing (weighted average), or Kalman filter technique. These techniques treat variations as trendless. The running average requires large data storage and gets nervous with few data points. The first-order smoothing technique requires less storage for data and is flexible in the weight applied. It is also limited to manual factor adjustment. The Kalman filter is more complex but adjusts gain downward as the forecast accuracy improves with experience.

Energy consumption rises as production hours rise. Second-order exponential smoothing is designed for forecasting this type of trend. Seasonality is also a factor in energy usage. Apply second-order smoothing to the production energy component and a sinusoidal function to the environmental component.

The Weibull function is particularly well suited to product testing but may also be applied to components subject to frequent failure. Its use requires over ten sample data points to achieve reasonable accuracy. This number of data points is frequently difficult to obtain.

MACHINE UPTIME

The goal in design, manufacturing, and plant engineering is to prevent production equipment downtime at the same time as doing other more specific functions. Redesigning products can reduce failures. Manufacturing engineering can plan methods and incentive systems to reduce equipment failures and increase productivity. Plant engineering must establish a good preventive maintenance system. Machines must operate reliably. Reliability must include product quality as well as uptime. Machines must produce a high-quality product and be available to manufacturing when it is needed to meet production schedules.

Machine failures fall into three categories:

1. Preventable
2. Predictable
3. Neither of the above

Preventable failures must be prevented. Predictable failures can be forecast using one of the techniques presented in this chapter. A good preventive maintenance system categorizes failures and works to reduce failure frequency and improve product quality. Forecasting is a weapon in the war on unexpected machine failures.

FAILURE ANALYSIS

Before prevention or prediction can be employed effectively, the mode of failure must be determined. Failures can be broken into roughly three categories:

1. Design defects
2. Processing and fabrication defects
3. Service deterioration

Design defects include sharp corners, wrong material, stress risers, improper fastening, inadequately computed loads, and inadequate stress analysis. Failures of this type are preventable. Fatigue failures of bolts or shafts with sharp fillets or keyways and fretting fatigue failures of press fits and bolted joints fit this category. Undersize power supplies, card edge connector burnout, chip failure due to inadequate isolation, and failures caused by relay contact resistance are the electronic equivalents of mechanical design defects.

Processing and fabrication defects include quench cracks, faulty heat treatment, forging or casting defects, and nonmetallic inclusions, to name a few. Misalignments, grinding cracks, welding flaws, and machining defects are examples of mechanical processing defects. Some of these defects can be located by nondestructive inspection. Magna-flux and magna-glo are the most commonly employed examples of nondestructive crack inspection.

Service deterioration is the third major cause of failure. These failures include stress corrosion, chemical corrosion, vibration, overload, and other environmental factors. No application is free of environmental hazards, but some of these factors may be compensated for. Fretting or stress corrosion cannot be eliminated, but proper lubricants applied on schedule reduce the risk of unexpected failure.

MECHANICAL FAILURE CLASSIFICATIONS

Design Defects

1. Errors in calculations of static tensile loads
2. Inadequate stress analysis applied to part shape
3. Brittle fracture from stress risers
4. Ductile failure from excess deformation
5. Fatigue failure
6. High temperature failure
7. Misapplication of material

1. This first item is a classic case of human error. Incorrect calculations are being eliminated by standard designs and computer processing but they still occur. Careful checking and rechecking by someone else helps avoid these mistakes.

2. Many failures come from high stress points due to part shape. Intersecting keyways and fillets create stress risers individually. Taken separately, neither produces a problem. Together they combine to produce fatigue cracks and brittle fracture. The key element here is the combination of shape features which causes failure. Failure analysis is a major contributor to the discovery of these failures. Some of the newly applied finite element modeling techniques on CAD/CAM systems model these factors very well.

3. Incorrect fillets, sharp keyways, and similar individual part features produce the third failure classification, stress-riser failure. These failures are similar to the second classification, but here a single feature provides the cause.

4. Ductile failures from excess deformation are design defects. As with the first classification, they are preventable with careful design.

5. Fatigue failures from load cycling, thermal cycling, fretting corrosion, and rolling element fatigue are the fourth major design defect factor.

The new computer modeling techniques are major contributors to eliminating these failures. Failure analysis was frequently the only way to detect these in the past.

6. High temperature failures from oxidation, local melting, creep, and warping are common design defects. Again, computers contribute thermal analysis and modeling techniques which permit detection of these problems at the design stage.

7. Misapplication of material is the last design defect classification. Caustic or hydrogen embrittlement, surface hard versus through-hardening, and dissimilar material corrosion are all examples in this group. Failure analysis is still the major method of detection of this type of failure.

Processing and Fabrication Defects

1. Material defects
2. Faulty material composition
3. Working defects
4. Machining and grinding defects
5. Faulty heat treating
6. Surface contamination
7. Welding defects
8. Improper assembly

1. Castings and ingot making are major sources of nonmetallic inclusions. Porosity, segregation, and pipes are other material defects that are introduced by processing.

2. Faulty material composition is a fabrication error. The correct alloying elements must be present in metallic materials. The correct chemical composition is necessary for plastic and other parts.

3. Working defects include delamination, seams, cracks, and laps. These are roll forming or extrusion-type defects.

4. Gouges, burns, and cracks are all machining and grinding defects. Some of these are detectable by inspection. Careful failure analysis determines the percentage of these in the total of all failures.

5. Faulty heat treatment includes quench cracking, grain growth, excess retained austenite, decarburization, and overheating. These lead to premature failures that are eliminated by process controls and process operator education.

6. Hydrogen embrittlement, chemical surface diffusion, plating, and coating all lead to surface contamination if they are done improperly. Process control is the solution.

7. Welding defects are eliminated by education and training. Porosity,

underbead cracking, undercuts, cracks, and residual stresses are generally avoided by well-trained personnel.

8. Bearing assembly into a housing with a hammer instead of an arbor press is an example of improper assembly. Others include dirt or foreign material inclusion, gouges, mismatched parts, and deformation from forced assembly. Careful process control can eliminate many of these.

Service Deterioration

1. Wear
2. Overload
3. Corrosion
4. Improper maintenance
5. Accidents

1. Wear is detectable by inspection. Erosion, galling, seizing, and the like are detectable before they become failures. Once failure occurs, they are predictable if design and materials are not changed when repairs are made.

2. Overloads caused by improper setups, out-of-spec. materials, and oversized parts require better process control and material quality control.

3. Corrosion is a fact of life. It can be reduced with coatings, different materials, and better lubrication. It will not be eliminated but life can be lengthened by corrective action.

4. Improper maintenance includes repair defects, an inadequate preventive maintenance program, and improper lubrication. An effective failure analysis can diagnose this problem.

5. Accidental conditions include severe vibrations, impacts, thermal shocks, and the like. When these are abnormal conditions and parts fail, design is not the issue. Earthquakes and other environmental conditions cause some of these. Furnace control failures cause thermal shocks and high-temperature failures. These type failures are unpredictable and unpreventable. Failure analysis prevents misdiagnosing failures in this category and including them in another category.

Failures can be made useful. They define the problems that need correcting. Good failure analysis is essential to an effective preventive maintenance program. Prevention is essential when possible, prediction is needed when feasible. A majority of failures are one or the other.

14

LONG-RANGE FACILITIES PLANNING

AIMS OF LONG-RANGE FACILITIES PLANNING

Determining the quantity and timing of manufacturing products really belongs in the domain of short-term manufacturing planning and control. By contrast, adequate long-term capacity depends on the timely acquisition of workers, material, equipment, and plant facilities, often a matter of lead times ranging from months to years. In the long-term process, however, it is important to use manufacturing control system data to determine the requirements for new facilities. This links space plans with business plans. By using data in long-range manufacturing plans, it is possible for engineering to plan for plant facilities procurement. Control actions insure in-time and within budget acquisition.

The data needed and the methods for facilities planning and control are available from the manufacturing control system. Engineering data are provided by properly designed manufacturing control systems. These systems need spatial data on file by department for both engineering and accounting to permit planning. Accounting needs spatial data for overhead allocation. Converting capacity planning to spatial needs via engineering methods, including space forecasting techniques, is a critical maneuver for both engineering and corporate management today. The forecasting techniques discussed in Chapter 13 are effective for space forecasting.

PLANT EXPANSIONS

Plant expansions are the direct result of the growing capacity needs of a company in the areas of manufacturing, office, and warehousing. Capacity limits for plants can be handled in much the same manner as any other capacity problem, with two important differences: 1. Plant construction lead times are months or years long as compared to days, weeks, or months necessary to purchase a piece of equipment; 2. the planning must include all phases and departments in a company, from accounting to sales. This has a company-wide impact as compared to the usual involvement of two or three departments in the purchase and justification of a piece of equipment. Plant expansions usually require large sums for capital investment and may entail moving whole departments or divisions thousands of miles. This process is much too difficult and complex, and too important to the overall success of the business, to do in response to crisis.

A facilities planning system must provide timely warning on the need for additional facilities that top management can accept as valid. With a company that has a manufacturing control or MRP II system, the basis for this timely, valid warning is resident within the long-range business plan. Although resident there, few have attempted to use this capability because it has only been available for a short time in a few companies with properly functioning control systems. Management—working with engineering and a system to provide the data—has a way to plan facilities to be available on schedule, within budget, and when needed to match management's business plan. The basis for this facilities plan is a space forecast which uses the business plan data.

SPACE FORECASTING

Forecasts provide the basis for most planning today, and facilities planning is no exception. It is useful to apply different schemes to make forecasts; these serve as checks upon one another and provide broader perspective. When available, historical growth data are useful to make forecasts of future needs. This type of future estimate assumes that the future looks like the past. Since most companies are in business to make the future different from the past—and hopefully better—this assumption is questionable. Historically based forecasts are useful as checks on other forecast systems because they are widely understood. The historical perspective is also useful if the forecast developed by another method system does not reflect past patterns. Deliberate measures change past practices so history won't be repeated. Historical forecasts measure the degree of optimism in other forecasts and detect overambitious goals.

All forecasts have the following basic characteristics stated previously in Chapter 13.

1. They are always wrong.
2. They should be stated as two numbers.
3. Family forecasts are more accurate.
4. The further a forecast extends into the future the more inaccurate it becomes.

Before choosing a forecasting system, evaluate some basic questions.

1. Is it reasonable to assume history will repeat itself?
2. Can the probable error be determined statistically?
3. Is the right item being forecast?
4. What variables produced the data?
5. Do these variables correlate with the data and reality?

Historical trends are useful so long as major business changes are not anticipated. Additions of new product lines, acquisition of new businesses, and obsolescence of an old product make history poor as a forecaster of the future.

The probable forecast errors determined by statistical methods are the most reliable if the future really is like the past. Be careful with statistics; the mathematics is highly precise but may have little validity. Chapter 13 covers additional details on forecasting.

The most difficult task is forecasting the right thing. The goal of long-range facilities planning is to determine space requirements, but the forecast is developed for capacity. A forecast for space develops from sales data. There is an indirect relationship between sales and space but a direct one between sales and capacity. The real relationship exists between productive capacity and space. It takes two machines to produce twice as much as one machine running equal times. This is modified if one machine is twice as fast as the other. Be careful to know the variables that produce the result in the real world. A forecast is a model. A model is a small imitation of the real thing. Be sure it fits reasonably well.

Be a skeptic. Does the forecast represent reality? Do forecasts for previous periods yield something close to the actuals in those periods? Forecasting space is no different from forecasting anything else; the result must correlate with reality. There are no pat answers to these questions; however, by reviewing them carefully prior to using the forecast for a specific purpose, planners can avoid serious mistakes.

HISTORICALLY BASED SPACE FORECASTING

Historical space data provide interesting analyses depending on which of three separate mathematical models is used.

1. A straight-line growth extrapolation by year and area. This is the typical linear historical growth in square footage by year.

2. A power curve growth to match more closely a compound interest growth rate. This is an attempt to project historical data on a curve rather than on a straight line.

3. A straight-line growth projection based on sales dollars per square foot by year. This is an attempt to forecast the sales dollars per square foot ratio over the past years and to use this ratio to forecast the square footage based on a sales forecast for the total company.

The first problem is which model to use. It is too complex and too confusing to deal with more than one historically based forecast, so a single model is chosen. Using a modern, small, programmable desktop calculator, one can easily evaluate the correlation coefficient for any mathematical curve fit. The correlation coefficient is a measure of the goodness of fit of a mathematical curve to a set of data points. The best fit is the method that has the coefficient closest to the value of 1.00. The formula for the correlation coefficient is:

$$r = \frac{n \, \Sigma xy \, - (\Sigma x)(\Sigma y)}{\sqrt{[n \, \Sigma x^2 \, - (\Sigma x)^2] \, [(n \, \Sigma y^2 \, - (\Sigma y)^2)]}}$$

Engineering has plant layouts which become the basis for square-footage measurements both for historical and capacity-tied forecasts. These archives contain information about the company available from no other source. Layouts provide the basis for the historical space analysis of a company. As additional facilities are required, engineering assembles the planning data with the assistance of several other departments in the company. Manufacturing control system data have not been available until recently for use in plant expansions in most companies. For example, the archive research might reveal historical data similar to that in Figure 14-1.

The sales data provide a measure of space productivity if inflation is removed from the sales volume and the sales dollars per square foot ratio is computed. The effective changes in sales per square foot with time provides the basis for another forecast. This multiplicity of potentially available bases for forecasting brings out the characteristics of all forecasts.

The coefficients for the three previous models for the data in Figure 14-1 have been computed. The low value of .9743 for the sales dollars per square

Figure 14-1.   ABC Company Ten Year Historical Data

Year	Square Footage	Sales in Millions
1969	975300	29.9
1972	1095100	36.5
1975	1257100	55.8
1977	1407100	73.7
1978	1463300	83.6
1979	1463300	92.5

foot ratio method proved this method the poorest. The coefficient of the power curve growth at .9971 was by far the most accurate mathematical fit to the historical data. The linear area by year model with a coefficient of .9836 lay between the power curve and the ratio model. Remember, the mathematics is precise but may not be accurate.

Graphically, the power function and linear fits are shown in Figure 14-2. Although the power function is mathematically a better fit to history, the following factors can modify the mathematical analysis.

1. Introduction and executive use of a manufacturing control system could reduce the future requirements for floor space at an ever-increasing rate.

2. Management would prefer to underbuild than risk overbuilding.

3. The lead times are such that the difference between the two methods are small enough to permit annual plant additions if a yearly review of space requirements indicated some needed action.

Figure 14-2. Power and Linear Curve Data Fit

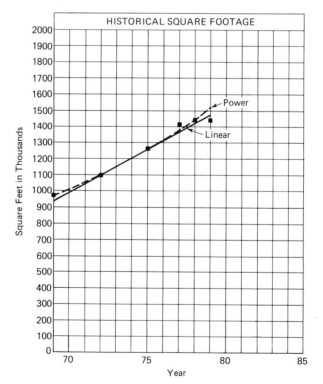

(Courtesy of George Plossl Educational Services, Inc.)

Table 14-1. ABC Company History-Based Forecast for 1980–1985

Year	Square Footage
1980	1568818
	1514672
1981	1624255
	1565186
1982	1679692
	1615701
1983	1735219
	1666215
1984	1790566
	1716730
1985	1846002
	1767244

4. The forecast error was in excess of the difference between the linear and power curve projection.

Thus judgment—and not the initial computer-generated numbers—provides the real key to the final forecast. Now the history-based forecast as defined is somewhat representative but isn't tied to the business plan or the data from the business plan.

The upper and lower limits, shown above in Table 14-1, are based upon a 68 percent probable upper and lower limit value based on mathematical probability analysis. This means that the real requirements in square footage would lie between the values indicated more than two-thirds of the time. This analysis was based on the variability of the past history data. Figure 14-3 shows a graphical representation of this forecast.

CAPACITY PLAN SPACE FORECAST

One company production planning group produces a five-year forecast of capacity needs by machine group. This five-year capacity plan provides the basis for a forecast of space needs based upon capacity. This forecast, compared to the historical, provides a link with the corporate manufacturing plan that the historical can't furnish alone.

To produce the space forecast, it is necessary to structure the capacity data in a manner compatible with spatial measurement. Engineering, using the plant layouts, assigns space to individual machine, office, and warehouse groups until all space is accounted for within the company's facilities. Production planning then provides the capacity utilization for each area in the engineering breakdown. The capacity utilization and maximum capacity for

Figure 14-3. Graphical Representation of Table 1 Data

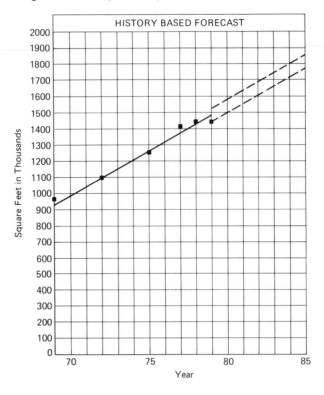

(Courtesy of George Plossl Educational Services, Inc.)

each of the years through each area is carefully worked out. Engineering evaluates the growth needs by area and develops a forecast of space needed by year based upon these capacity data.

This process was used to forecast space needed for the company's expansion. The expansion resulted in the construction of a 146,000 square-foot building at one plant location. The capacity space forecasting system used the capacity plan growth percentages by major machine group to project future needs based upon present space usage. This meant that if it took 1000 ft^2 to house a machine at present, it would require 1100 ft^2 to house a 10 percent growth in capacity. This assumption produced some difficulties when applied to underloaded work centers and did not account for the historical improvements in space efficiency. Plants improve in space efficiency by producing more product per square foot. Nor did this forecasting system account for machines having to be purchased in units of one machine, which required stepped growth.

The assumption was made that, in the total forecast, the effects of small percentage additions to numerous machine groups would produce the same result numerically as the addition of fixed units of one machine, an assumption that was correct for this particular company. They had hundreds of machines and they were treated as a entire group.

A correction factor is applied to compensate for the historical improvement in space efficiency of manufacturing area. Again, the archives reveal the per square foot improvements in efficiency. For the company cited, in the past ten years the output of manufacturing rose nearly 40 percent per square foot. The company produced 40 percent more in the same area by means of increases in numbers and kinds of high-speed equipment, better methods, and better use of work centers. The space efficiency improvement correction factor was 4 percent per year. This would mean that the 1000 square foot group mentioned before would be forecast at 1060 square feet. The capacity growth increased the space by 10 percent, and the efficiency factor reduced the gain by 4 percent. This is the method to use on a work-center-by-work-center basis for almost any large company. For small companies, the increases in capacity can be applied to individual machines more easily. The result is the same—a capacity-based space forecast.

SPACE ALLOCATION
FOR THE CAPACITY-BASED FORECAST

To apply the methods outlined previously, it is necessary to build an allocation system to account for the total space in the company. This allocation requirement presents a number of questions. What area is really required for a machine? What are the real work-in-process space needs? How should aisles and common areas be allocated? What are finished goods on the manufacturing floor—manufacturing or warehousing? Where should raw materials stores be allocated? To achieve some consistency in the approach to allocation, the following guidelines to space categories may be helpful:

1. Warehouse space is finished goods storage space in a warehouse environment. Areas containing finished goods staged to move to the warehouse are in manufacturing space.

2. Office space is true office areas occupied by people serving many departments. This includes general offices, quality control, production control, and centralized engineering. Supervisors' offices and toolrooms are part of manufacturing.

3. Manufacturing space is that area occupied in and around production equipment, including aisles, raw materials storage, work-in-process stores, and other areas necessary to the production of the product.

4. Miscellaneous space is the catch-all for vaults, first aid, and other areas required by the plant but not really belonging to one of the other categories.

Given these definitions, engineering analyzes the floor layouts and calculates square footages for each machine group and area within the confines of the building walls.

Figure 14-4 shows a typical space allocation format. For multi-plant companies, each machine group can be split by geographical location. This produces a summary sheet like the one shown in Figure 14-5, which takes each building area floor by floor and allocates space within the company. This procedure provides an accounting of square footage that makes it considerably more difficult to lose any area that may be buried in some corner. Given the specific growth percentages from planning for each machine area, it is now possible to generate the forecast.

Figure 14-6 shows the capacity plan forecast for the period 1980–85 for the sample company. Since the process is a mathematical one, it yields a single value rather than two values required in a forecast. Since it is rational to assume that the range seen in historical values is representative, ranges for the capacity plan–based forecast can use the 68 percent probable values from this model. Figure 14-7 shows the historical and the capacity plan forecast plotted

Figure 14-4. Typical Space Allocation Format

TYPICAL SPACE ALLOCATION FORMAT

DIVISION

Growth	Machine	Location	Location	Location	Location
%	Group	A	B	C	D
5 Year		Sq Ft	Sq Ft	Sq Ft	Sq Ft

Total Division

(Courtesy of George Plossl Educational Services, Inc.)

Figure 14-5. Space Summary Sheet

TYPICAL SPACE ALLOCATION FORMAT

SPACE SUMMARY

Building Area	Square Footage	Division			Office, Whse., Other

Totals

(Courtesy of George Plossl Educational Services, Inc.)

for comparison. Note that the square footage projected by the capacity plan is significantly greater than that forecast by the historical method.

This brings us right back to the philosophical questions asked in the early stages of development of the space forecasting system. It also leads us back

Figure 14-6. Capacity Plan Forecast

1980 - 1985 SPACE FORECAST

CAPACITY-PLAN-BASED

Year	Square Footage
1980 — — — — — — — — — — — —	1,578,300 1,524,100
1981 — — — — — — — — — — — —	1,671,700 1,612,600
1982 — — — — — — — — — — — —	1,765,100 1,701,100
1983 — — — — — — — — — — — —	1,858,500 1,789,500
1984 — — — — — — — — — — — —	1,951,900 1,878,100
1985 — — — — — — — — — — — —	2,027,100 1,948,400

Figure 14-7. Capacity and Historical Forecast

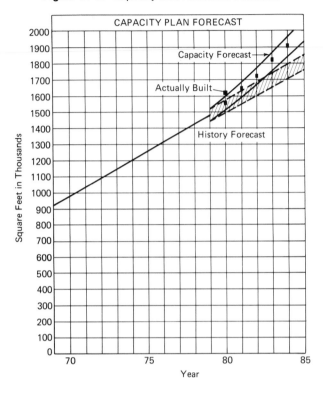

(Courtesy of George Plossl Educational Services, Inc.)

to the reason the linear projection of square footage by year was chosen as the history-based forecast model. Judgment validates a forecast. For the company data already presented, the 1981 requirement of 1,642,000 ft^2 was an addition to the facilities of 182,000 ft^2 over the former 1979 total. A convenient addition fitting the property and present building would probably satisfy those requirements.

An expansion of 150,000 ft^2 would cost as much as five million dollars. With adequate planning it is possible to develop building plans and specifications that will keep costs under this amount, reduce overall energy consumption in the long term, and provide the needed space. The manufacturing control system supplied the data necessary to develop spatial requirements to meet the capacity plan. We can now use the manufacturing control system data to determine the type of space. Building design and specification vary greatly depending upon the mix of manufacturing, warehouse, or office requirements.

PLANT LAYOUT BLOCK PLANNING

The basic raw allocation data and the growth percentages provide the total requirements in square feet through 1984. Again, we can go back to our forecasts and allocation data to break apart the specific needs for space into manufacturing, warehouse, office, and miscellaneous. This provides the space requirements by year for the period from 1979 through 1982 by type of space. The 1981 needs were for 42,000 ft^2 of manufacturing, 150,000 ft^2 of warehouse, 5000 ft^2 of office, and 20,000 ft^2 of miscellaneous. Since 63 percent of the space required was for warehousing, it looked like the majority of the expansion would be warehouse. The fact that 1982 would increase the space requirements for warehouse to 173,000 ft^2 only reinforced that position.

For example, if a particular plant location has the following characteristics, we can apply expansion data to this facility.

1.	Total space	130,510
2.	Manufacturing space	83,536
3.	Warehouse space	40,000
4.	Office space	4,200
5.	Miscellaneous space	2,774
6.	Bay size	34 ft x 34 ft

Assume that the utilities, heating, ventilation and air conditioning system, and floors are designed to support manufacturing in this plant. By moving manufacturing into the area occupied by the warehouse in the present facility, the manufacturing needs for 1981 could be met very closely. This indicates that a warehouse is the most useful approach to this expansion. Although these conditions may appear artificial, similar logic can be applied to any plant expansion.

Energy is a major factor in this expansion. To reduce the energy consumption of the building, extra insulation is put in the walls by foam-filling the wall blocks and increasing the thickness of the fiberglass insulation in both the metal-insulated panel walls between the block and the roof and in the roof itself. The heating, ventilating, and air conditioning systems will be multiple rooftop industrial units each individually controlled with its own zone thermostat; a mercury capsule fixed-temperature type will virtually eliminate tampering. All rooftop units for both buildings, the old and the new expansion, will be controlled by a central demand and load scheduling computer which will shed electrical loads during peak electrical periods and will cycle air conditioning units off or on to night setback during periods of nonoperation. The lights will be located wherever most convenient in the structure by incorporating a 100 foot-candle power grid. This grid will permit lights as required to illuminate the desired areas adequately. All truck dock doors will be equipped with dock seals to limit infiltration during loading and unloading of tractor trailers. These energy conservation measures will be easy to incorporate at the building design stage and are well worth the investment in today's high energy cost environment.

The cost estimate for the final expansion size of 146,000 ft^2 can now be made because the structure is defined. The warehouse engineering for rack selection, item location, purchased rack installation, and operation can now be done by engineering.

Similar analyses can be made for any corporation if the historical data are available. Old layouts on CAD systems are ideal for historical analysis. Capacity planning–based forecasting, modified by the historical approach, is considerably better than a historical approach alone. The data assembled for the capacity-based forecast is used later and has more validity if the long-range business plan is realistic.

15

THE JUST-IN-TIME PHILOSOPHY

Just-in-time is not a Japanese or other foreign concept. It has a very simple definition and doesn't depend upon culture, nationality, or type of business for success. Just-in-time is a strategy with specific tactics.

DEFINITION

Just-in-time is a strategy for achieving significant, continuous improvement in performance by the elimination of all waste of time and resources in the total business process.

JIT requires and stimulates simultaneously:

Improvement of quality in all activities
Reduction of lead times
Improvement of productivity
Release of the latent capabilities of all people in these improvements
Improvement and extension of quality, lead times, and productivity to supplier and customer networks

Just-in-time is a total business strategy. It will not be nearly as successful if applied in single departments. Departments are interrelated and strategic plans for one may conflict with the plans for others. Only as an overall strategy is just-in-time truly effective.

JUST-IN-TIME TACTICS

JIT tactics include a frontal assault on total business cycle time and work-in-process by reducing lead-time length, lead-time variability, lot sizes, and setup times. Lead-time reduction requires quality improvement, and quality improvements are by-products of process improvements. Process monitoring and improvement eliminates disturbances, thus causing overall quality to improve as the process gets more reliable. As process reliability improves, product quality stabilizes. Stable product quality reduces interruptions and disturbances in smooth product flow, thus reducing lead times and costs through reductions in scrap and rework. The continuous pursuit of process reliability and quality are universal tactics in a just-in-time strategy.

Henry Ford used just-in-time when the Detroit, River Rouge, Ford plant was built and operated in the early 1900s. Iron ore from barges was converted to automobiles within 48 hours of receipt. This required that all processes be ready to operate and that nothing sat on the shop floor awaiting action. Processes ran continuously and material kept flowing. Lead times were short, and great effort was expended to reduce them further.

Efficiency is the ratio of output to input. The output of product per hour of production is production efficiency or productivity. Anything which delays production, interrupts production work or material flow, or doesn't contribute to production lowers efficiency. Anything which lowers efficiency is waste.

Another tactic in the just-in-time strategy is a war on waste. Any operation, person, or material problem which hinders smooth product flow is improved or eliminated. Operations, departments, and activities add value to the product or they waste resources needed to add value to the product. JIT is not a typical "defect reduction program," but it does attempt to reduce defects by improving process control.

Everyone in the manufacturing organization contributes to overall productivity. The JIT strategy enlists everyone in productivity improvement. JIT recognizes that maximum productivity depends on maximizing asset usage, and human resources are a company's most important asset. Therefore, this strategy attempts to release the latent capacity of each employee to attack productivity. All personnel are expected to contribute to improving product quality, reducing lead times, and eliminating waste.

Since customers and suppliers have great influence on manufacturing flow and productivity, JIT strategy develops tactics to enlist them in the job as well. Suppliers are encouraged to deliver on time, in small quantities, and with zero defects. Customers are encouraged to order in small quantities to meet short-term needs and to share long-range forecasts of usage rates. Communications with both suppliers and vendors must improve to have short lead times and high service levels.

WHY JIT?

The pressures for moving toward lower inventories and just-in-time manufacturing philosophy are primarily from international competition. Figure 15-1 shows inventory levels and inventory turns achieved by this competition. The pressures have never been greater for achieving real control of manufacturing. The reasons for not achieving better control are numerous but some need mentioning.

1. The historical emphasis has been to keep capacity fully utilized. This is a direct result of attempting to maximize operation of expensive labor and equipment.
2. The fact that maintaining high utilization resulted in producing excessive inventories was previously ignored. Accounting labeled inventory an asset and assets were good to have because they increased net worth. This ignored the fact that inventory is not an asset but a liability when resources of labor or material are scarce and the processes needlessly consume both.
3. Direct labor costs have historically been the focus of management. This may be the result of the time when the greatest percentage of the cost of producing a product was direct labor. With the ever-rising costs of materials and production capacity in equipment, this emphasis is misplaced now.
4. Large lot sizes are good because costs of setups amortize over many pieces, reducing the piece part cost. It is also easier to schedule one long run than many short ones just in terms of paperwork.

Figure 15-1. Inventory Levels and Turns

INVENTORY LEVELS AND TURNS

Company	Inventory	Turns
Toyota Motors	3 Days	120+
Kanto Auto Works	3 Days	120+
Nissan Motors	3 Days	120+
Nippon Denso	5.9 Days	60+
Kawasaki	4.5 Days	80+
Sony	21 Days	17
Tomy Toys	7 Days	52
Arai-Garlock	18 Days	20

5. Material tracking, material handling, and labor reporting are complicated by short runs, and thus long runs have been favored by management.

6. Materials management technology has focused on just-in-case inventory to bury problems with cushions of inventory. This is inventory theory A logic, which we will cover shortly.

7. Traditionally, management has sought solutions to inventory problems in complex computerized systems that merely work around the physical realities of the manufacturing plant.

8. Consultants and experts in materials management have become so enamored with buzzwords and computerized techniques that the need to solve real problems has been largely ignored.

Inventory theory A states that downtime is intolerable from material shortages. Wide acceptance of this theory is confirmed by just-in-case actions which support this theory. Raw materials are kept in stock to prevent downtime when vendors don't deliver and because materials are defective. Work-in-process is high so machines don't run out of work, material record errors can be discovered, and the job can be called in-process. Finished goods warehouses are kept stocked so customers can be supplied quickly and demand variations can be compensated for, and as a cushion against labor unrest or other interruptions in supply.

Materials are retained in inventory to prevent stockouts and production downtime. Inventory is kept like insurance. It cushions against unexpected interruptions or fluctuations. Like insurance, however, there is a price to be paid. Figure 15-2 is a diagram of the movement of a stocked item over time with reorders computed using the order point technique.

The average inventory quantity is 15 over the time horizon. The cost of that average inventory is 15 times the unit value. Thus for an item with a $10 unit value, the average inventory is $150. At a 30 percent carrying cost percentage, this inventory level cost is $50.00 per year. If the usage rate of five per week could vary to 10 per week and orders can be received a week late, the safety stock to cushion against these interruptions raises the average inventory level to $250. This has a cost of $75.00 per year, an insurance premium increase of $25.00 per year, or 150 percent of average inventory cost for this item. Sum the costs of carrying inventory for each item for the hundreds of items in a manufacturing plant and the total is a significant percentage of the earnings.

MRP II or manufacturing control systems reduce these costs. These systems anticipate usage for dependent demand items and they work well to minimize inventory. Figure 15-3 shows the effect of the manufacturing control or MRP II system on the manufacturing plant.

Given a specific lead-time length, the system attempts to balance flow volume and plan priorities. The company desires the right total quantity to

Figure 15-2. Order Point Material Flow

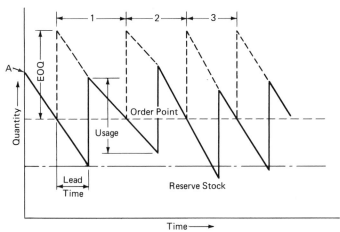

(Courtesy of George Plossl Educational Services, Inc.)

Figure 15-3. Manufacturing Control or MRP II Control

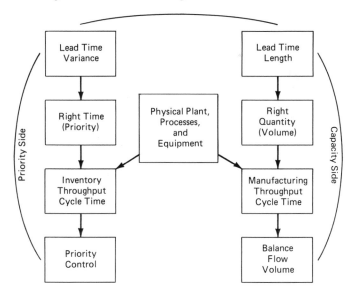

(Courtesy of George Plossl Educational Services, Inc.)

satisfy customers and a priority which satisfies those customers at the right time. The planning and balancing functions are made complex by the product structure, physical plant processes, and equipment. The manufacturing segment cycle time is the time it takes to make products within the factory. The inventory segment cycle time is the time it takes to move inventory through raw materials, work-in-process, and finished goods to the customer. The manufacturing and inventory cycle times result from plant environmental characteristics and plant methods.

Manufacturing control or MRP II systems attempt to plan plant operations around existing plant conditions. Planning software will not change those conditions. To significantly improve plant performance and the ability to run the plant, the resource characteristics and methods must change. Some of these plant environmental characteristics are listed in Figure 15-4.

Customer and vendor relationships are directly analogous. Marketing functions as the eyes and ears of the manufacturer in the marketplace of its customers. Purchasing functions as the eyes and ears of the manufacturer in the marketplace of its vendors. Smooth outflow of product to customers requires smooth inflow of raw materials from suppliers to minimize resource waste. If marketing and sales are "order takers" and oblivious to leveling and balancing the inflow and outflow of work, this contributes to surging workloads. This makes planning and control more difficult and true JIT impossible.

Purchasing procedures such as, "Nobody talks to vendors except us," complicate manufacturing planning and control. The six purchasing rights are:

> The right material
> The right time
> The right quantity
> The right quality
> The right price
> The right vendor

Figure 15-4. Plant Environmental Characteristics

> Customer relationships
> Vendor relationships
> Purchasing procedures
> Product varieties
> Product designs
> Facilities layouts
> Processes
> Quality control or management
> Toolroom operations
> Setup methods and controls
> Material handling
> Maintenance procedures

Only two of the six purchasing "rights" (or correct actions) are solely purchasing activities. Only vendor and price are contributed directly by purchasing. Material comes from engineering, time and quantity from material control, and quality from engineering and quality assurance. Purchasing relays messages from these other functions to vendors. These other activities must communicate with vendors to balance real needs with supplier capabilities. Anyone who has played the game "telephone" as a child readily recognizes that messages get distorted with each person-to-person relay. Good communications and smooth material flow require direct contact between vendors and material control, engineering, and quality assurance. Purchasing can simplify planning and control by smoothing out raw material and component inflow and fostering effective communications. Sourcing is vital and purchasing must do that job effectively.

Product varieties and designs assist or hinder manufacturing. A company that assembles to order thousands of end items from combinations of a few standard modules simplifies planning and control and makes it far more effective. Designs developed independently from individual components instead of modules are more complex. The degree to which engineering is designing products using modules is a large factor in manufacturing control.

Facilities layouts are partially responsible for smooth or erratic material flow. The more convoluted the flow path from work center to work center, the more erratic and surging the flow. No material control system can hope to improve flow beyond the capabilities of the plant layout to physically permit it to move.

Processes and effective process control are essential to short lead times. Outmoded equipment with little process control cannot hope to produce reliable quality. This doesn't imply that new is better, but that equipment with effective process control is newer. Neither does it imply that new high-speed, long setup equipment is better. Old designs for simple machines that are easy to set up and control produce quality products reliably. The ultimate sophistication is simplicity.

Current emphasis on quality control misses the issue completely. Quality can be controlled but it should be managed. Quality management carries the idea that quality can be continuously improved by effective management. Quality control implies that there is some finite quality level that represents an acceptable level of defects. Statistical methods compute Average Outgoing Quality Levels (AOQL) and quality control programs determine the acceptable AOQL. Acceptable for today but not forever is implied by quality management.

Toolroom operations have a dramatic effect on manufacturing cycle times. The scheduling of tool repair, the commonality of shut heights, and the setup time required to put the tools into service all contribute to manufacturing cycle times. Setup methods and controls within manufacturing establish manufacturing cycle times and lead times.

Material handling is a misnomer for many companies. Material handling implies material movement, but many material handling studies are directed toward finding more cost-effective ways to store the material, not move it.

Maintenance procedures and preventive maintenance programs directly affect quality and process reliability. No computerized system can hope to compensate for unreliable processes.

These are some of the environmental factors that the manufacturing control (MRP II) system is asked to plan for. The more complex and unpredictable the environment, the more complex and unreliable the planning and control system needed to cope with it. Figure 15-5 likens the manufacturing control system to a system to control a ship in a river. If the water level is raised (increase inventory), the controls are not needed with the problems covered. Reducing water level (inventory) uncovers islands (problems) and complex controls are required.

Just-in-time strategy uses tactics which simplify and make the environment more reliable. Manufacturing control (MRP II) systems don't compete with Just-In-Time. The JIT tactics simplify planning and control by improving the physical manufacturing flow. This is engineering the total manufacturing environment.

Figure 15-5. Complex Controls

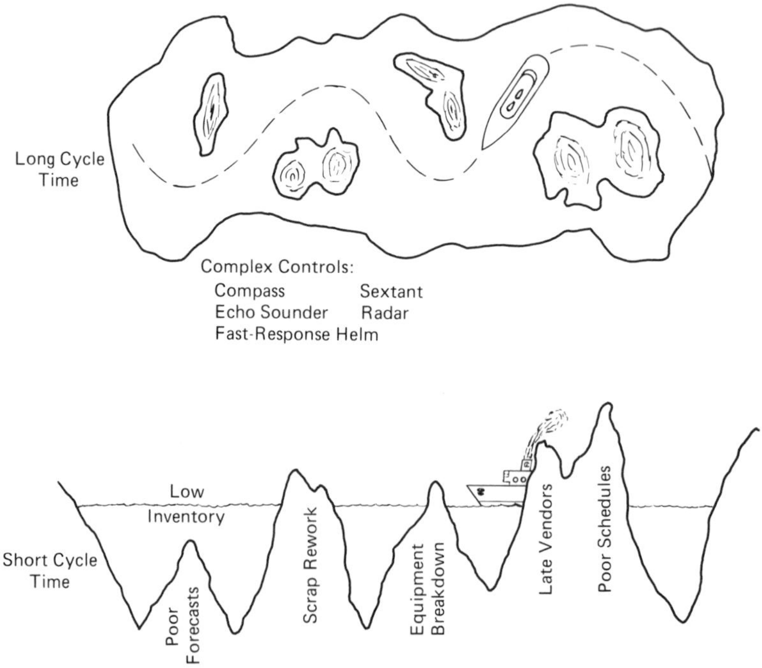

(Courtesy of George Plossl Educational Services, Inc.)

ENGINEERING THE TOTAL PROCESS

Total process engineering starts with acceptance of inventory theory B. Inventory theory B states that inventory masks certain problems that must be solved. Although it is impossible to eliminate all problems, JIT strategy requires an unrelenting assault on all that exist.

Theory B also eliminates EOQs with a massive assault on setup time. Single-minute exchange of dies is a hallmark of this theory. The concept is to make the first part as easily and cheaply as the hundredth and to make only what customers are ordering.

To make only what customers order demands the third item in theory B—flexible capacity. Flexibility is achieved by cross-training people, acquiring general- not specific-purpose machines, and rearranging plant layouts to keep pace with routing changes to improve productivity. Plant capacity must be flexible to respond to changes in customer demand. Capacity must be available to match customer requirements and ordering patterns. When capacity is available and flexible, lead times can be short.

Theory B requires driving lead times toward zero. Real customer service is the ability to supply what the customer requires each day. JIT and theory B work toward shipping from manufacturing today what the customer orders today. This requires flowing the product, not stocking it. This requires matching demand rates and production rates.

Figure 15-6. Manufacturing Control with Just-In-Time

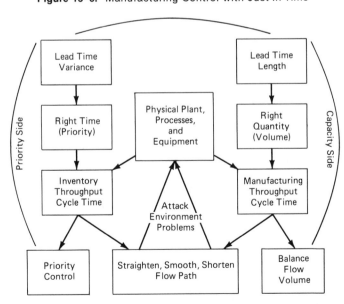

(Courtesy of George Plossl Educational Services, Inc.)

In short, inventory theory B states that inventory is evil and must be eliminated. To engineer the total process, theory A must be discarded in favor of theory B. Theory A covers "unavoidable" problems with inventory. Theory B eliminates disruptions and flows the product. Theory A assumes that problems are more easily covered than solved. Theory B views problems as waste which must be eliminated. It is amazing to see how many unavoidable disruptions are eliminated when effort is expended under theory B. The total process can be engineered. Figure 15-6 shows the relationship of JIT and inventory theory B to an MRP II system.

Engineering the total process requires understanding that software-based material control systems attempt to plan given existing plant conditions. These conditions are the sum of the critical path activities required to produce and ship a product. To improve performance, manage the total manufacturing process and the factors that affect it.

THE IMPACT OF JIT ON ENGINEERING

Major changes must occur in the traditional engineering environment. Emphasis shifts from minimum cost to adequate cost for manufacturable designs. This requires improved communications among engineering and manufacturing people. Design engineering works with manufacturing and production engineering on designs in a JIT environment.

Tools and equipment are designed and selected with a quick setup emphasis. Tool and equipment life must be long and operations must be reliably maintained. Plant engineering works with production and manufacturing engineering to keep the plant producing quality products.

Quality must be designed and built into the product. Designs which are manufacturable within production tolerances have quality built in. Processes don't produce bad parts to move to subsequent operations; instead they are fixed. Quality monitoring is everyone's business in a JIT environment. Engineering is under constant pressure to improve quality by improving processes and process control. Design engineering is called upon to assist manufacturing engineering, and production improves quality by design/process optimization.

The emphasis in the entire manufacturing plant and offices shifts from detecting problems to solving them in a JIT environment. Finding problems is easy—fixing them is difficult. Solutions for JIT are solutions which improve the whole. Problems are not solved in one department only to create new problems elsewhere. The whole organization, from the highest management level to the newest unskilled employee, is enlisted in solving problems. Problem solving is unrelenting because new and different problems always exist.

JIT is a strategy which requires dedication to perfection at all levels in manufacturing. It is difficult to achieve. It takes a long time and an unswerving purpose to make significant gains. JIT is not a short-term program or project, but a lifetime avocation. It is not a goal, but a journey. It is not a collection of techniques and it is not limited to one group, department, or to suppliers. JIT is not a competitor of manufacturing control (MRP II) or any other computer-based system; it improves the functioning of planning systems by improving execution. JIT is a total business strategy.

THE TEAM APPROACH

The sequential project concept shown in Figure 15-7 is typical for many manufacturing companies. Each function is compartmentalized and suboptimized independently. Information flows from one area to another in a formal, rigid path. Each operation or department completes its function before another starts. This rigid sequential system leads to manufacturing rigor mortis. Inflexibility in responding to changing conditions and rigid rules inhibits the flow of information needed to improve the whole.

ENGINEERING IN THE TWENTY-FIRST CENTURY

Figure 15-8 illustrates the integrated project concept needed by the twenty-first-century manufacturing company. This concept views the whole, as a small company, integrated in the total design–development–production process. This is matrix management—the team approach.

Computer Integrated Manufacturing (CIM) will become a reality. CIM will start with the design group being able to get data on processes from an integrated data base. Design will work cooperatively with manufacturing to develop manufacturable designs. This is effective CAD. Computers will produce NC machine data (tapes, programs, etc.) directly from design data; this is true CAM. Computers will assist with manufacturing process planning (CAPP). Capacity requirements planning (CRP) data will help develop manufacturing strategies. It will use group technology concepts and will provide operation simulation capabilities. The manufacturing control system (MRP II) will tie financial and manufacturing planning and control together with engineering and JIT to produce integrated manufacturing control. This is computer-integrated manufacturing, a concept that will make the twenty-first-century manufacturing plant operate like a single machine.

BENEFITS TO INTEGRATION

The benefits to engineering are numerous. Direct cost reductions are possible all along the way. Shorter design times by ten to 20 percent result from standardization, easy data entry, and fewer interruptions. Parts standardization

Figure 15-7. Sequential Project Concept

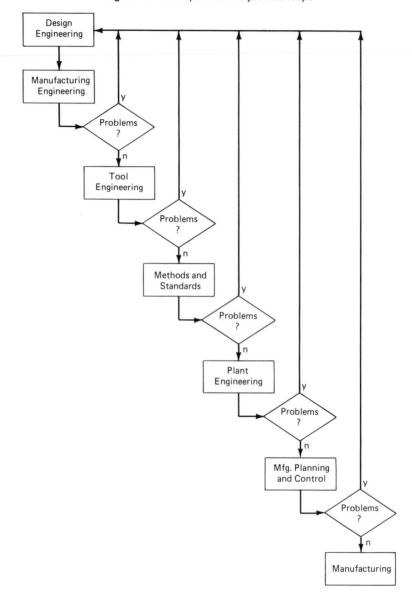

(Courtesy of George Plossl Educational Services, Inc.)

and optional assembly of modules for end item customization requires design engineering creativity. Easy data entry to computers with CAD and the integration of CAM makes products more manufacturable. Fewer interruptions occur with more manufacturable products.

Figure 15-8. Integrated Project Concept

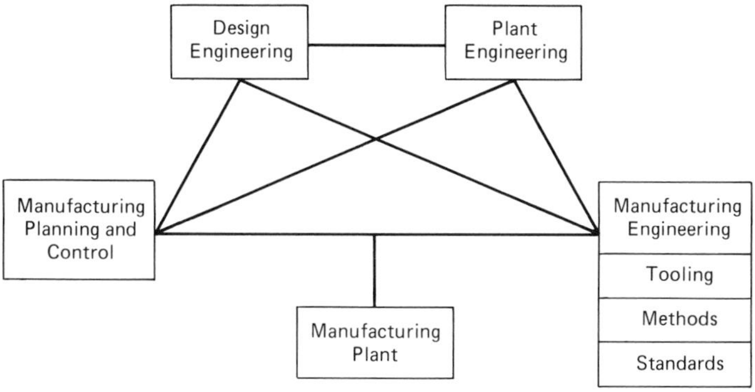

(Courtesy of George Plossl Educational Services, Inc.)

Drafting time is reduced by 20 to 40 percent because of CAD and the bill of materials processor in the computer. The bill of materials is more accurate and easier to check because the CAD computer and the business and material planning computer share data. Assembly drawings are produced by using the bill of materials in the business machine as input to the CAD system for parts required. The CAD computer draws the parts based upon part numbers and engineering drawings for the parts. An inaccurate bill is immediately apparent in this environment. Smaller drawings are a by-product. It is no longer necessary to put the bill of materials on the drawing. The business computer has the bill data and CAD links the drawings. The print file is necessary only for obsolete parts. Current drawings can be made as needed, and computer-generated drawings can be put on CRTs in the shop if desired. For design engineering, this reduces engineering cycle time and reduces the difficulty of making engineering changes. The greatest agony involved in making engineering changes is determining when to make them. A smoother flowing production operation makes this easier.

Manufacturing engineering also benefits from this simplification. Standards and methods stored in computer files instead of paper ones yield 75 to 90 percent reductions in file maintenance labor. The shop is better serviced because personnel obtain current information directly from the computer. This releases engineering time to work with design and manufacturing to make the product more efficiently. Processing time is shortened by 25 to 40 percent from design and routing standardization. Improved data entry and lessened paperwork release time for JIT studies.

Improved control of maintenance and repair supplies, tools, and equipment spares are benefits to plant engineering. When products flow through the shop and the computer schedules preventive maintenance as well as production, these improvements are significant (20 to 40 percent). The computer handles inventories of parts and supplies, work orders for repairs, and costing. This releases time for effective energy planning and control and long-range facilities planning. Energy costs drop 25 to 50 percent in plants with integrated systems.

Effective preventive maintenance, proper priorities, available spares, and correct tools all work to reduce the maintenance load by 10 to 30 percent. Machine uptime takes precedence over breakdown repair time. This emphasis shift reduces downtime permanently, with continuous machine improvements. Priorities for machine repairs get the vital few machines fixed correctly and quickly. Better coordination with manufacturing and design engineering get machines that have more common spare parts in stock. This makes spares more available and easier to obtain.

JUST-IN-TIME AND SUCCESSFUL COMPANIES

Successful companies don't believe that problems must exist—they view them as intolerable and go about solving them. They don't believe they have the ability to bury money in inventory; they put it in the process to produce products. They believe that the capacity to produce is a valuable asset and customer service is greatest when product flows. They believe in the effective use of all people's talents. The customer and the producer are both served by shipping the product ordered today directly from production. Successful companies believe that good management is the long-term guidance of all assets to maximize quality, customer service, and efficiency. They believe that the rewards of doing this well are profits.

These beliefs are those of just-in-time production. Successful companies have proven the validity of these concepts for many years in many countries in every industry. When management loses sight of the fundamentals, the company is jeopardized. Engineering needs integration, not specialization, for successful manufacturing. This is engineering for control of manufacturing.

PROBLEMS

Chapter 1 Engineering in a Manufacturing Company

1-1. Visit a company or study yours and draw an organizational chart showing the relationships between engineering and other departments.

1-2. What department is responsible for part numbers?

1-3. Name three benefits of quick-change tooling.

1-4. Discuss the reasons for conflicts between:
   a. Design and manufacturing engineering
   b. Manufacturing and plant engineering
   c. Engineering and manufacturing

1-5. Why is plant engineering normally assigned energy control and energy planning?

1-6. Explain the reasons for teaching computer programming in engineering school today.

1-7. What are some of the major problems in manufacturing with today's computer systems?

1-8. What is MRP logic applied to projects called?

1-9. Why is functional integration so important today?

1-10. List ten functions needing integration to improve manufacturing.

Chapter 2 The Manufacturing Control System

2-1. Identify the principal interests of engineering, manufacturing, sales, finance, and top management in manufacturing control systems.

2–2. List ten reasons why a manufacturing control system can fail to perform properly.

2–3. What are the basic requirements for effective planning and control? Discuss each.

2–4. Name some of the enhancements to the square root EOQ formula.

2–5. Is the MRP technique open or closed loop? Explain.

2–6. Discuss the reasons for the development of the manufacturing control system.

2–7. What are the functions of the MPS?

2–8. Define planning, control, and execution. Show how these activities relate to improving manufacturing.

2–9. Why is the antenna in the MPS shown in Figure 2-4 produced in a pattern 0-1-2-0-1-2 . . . instead of 1-1-1-1-1-1 . . . ?

2–10. What is a bill of labor?

2–11. How is a bill of labor used to verify the master schedule?

2–12. Explain the effects of changes in setup time and improved flow on the bill of labor.

2–13. What are the three functions of material requirements planning?

2–14. Redraw Figure 2-8 using a lead time of less than a week between levels in the bill of materials for manufactured items. Use the lead time shown for the purchased components.
   a. Which week number has the requirements for the 100 purchased components?
   b. What is the total lead time needed to make final assemblies?
   c. What industry can you name that has manufacturing lead times of under a week?

2–15. What are the four functional groups which comprise a complete manufacturing control system?

Chapter 3   Design Engineering

3–1. Discuss the three classifications of technical creativity.

3–2. Name five traits of creative people and explain each.

3–3. In which phase of the design process is the bill of materials developed?

3–4. What are the four types of organizational structures? Discuss each briefly.

3–5. Which type of organizational structure does the company in Problem 1–1 have?

3–6. What is the real goal of design engineering? Describe the reasons for this goal.

3–7. Discuss the value of plant routings to the design engineer.

3–8. Discuss the value of more manufacturable designs to manufacturing and the effects on the material control system.

3–9. Make a list of the data in the manufacturing control system useful to:
   a. Design engineering
   b. Manufacturing engineering
   c. Plant engineering

3–10. Where does the integration of functions in manufacturing start?

Chapter 4   Computer Systems

4-1. Name the types of computers you have available to you. Discuss each briefly.

4-2. For one of the computers in Problem 4-1, what type (8, 16, 32 byte), speed, and other characteristics does it have? Does it have a hard disk? If yes, what size?

4-3. What types of memory does your company or school computer have?

4-4. What peripheral devices are connected to it?

4-5. Why must color graphics be done on raster tubes today?

4-6. What tube type has the best graphics image quality? Why?

4-7. What programming languages does your company or school computer have available?

4-8. Discuss the value of improved productivity in engineering from computer use.

4-9. List and discuss some of the benefits to management of computerizing engineering.

4-10. Visit a typical CAD/CAM system installation. Draw a configuration diagram like the one in Figure 4-4 for that system. Is it tied to a business mainframe?

4-11. What is the effect of doubling the cost of the hardware and software to double the productivity of a typical CAD/CAM system (all else being equal)?

Chapter 5   Computer Programming

5-1. What do programs consist of?

5-2. What does 135 in base 10 equal in hexadecimal?

5-3. Explain why hexadecimal programs were easier for programmers than those written in machine language.

5-4. Discuss the differences and advantages of assembly language over machine language programs.

5-5. What types of programs are best written in assembly language?

5-6. Discuss the advantages of high-level languages.

5-7. Discuss the reasons for language and computer incompatibility.

5-8. What are the three advantages to commercial software? Discuss each.

5-9. What software module is most frequently custom written in a manufacturing control system? Why?

5-10. Discuss the advantages and disadvantages of in-house program development.

5-11. Explain how batch manufacturing control software differs from repetitive or just-in-time software.

5-12. For your company or one you visit, what software languages are currently in use?

Chapter 6   Computer System Interfaces

6-1. What are the two types of communications interfaces?

6-2. What are the advantages of each type?

6–3. Why is synchronous data transfer faster than asynchronous?

6–4. Discuss the advantages of synchronous and asynchronous data transfers.

6–5. Discuss which system you feel would be best for interfacing a CAD/CAM and business mainframe.

6–6. Discuss the characteristics of the three major network configurations.

6–7. What is the main advantage of the ring configuration?

6–8. Explain which of the three levels of CAD/CAM–mainframe you feel would be the best for your company or a company you have visited.

6–9. What does the term *signal conditioning* mean when applied to process control systems?

6–10. What is the computer language most frequently used in real-time process controllers? Why?

6–11. What are the advantages and disadvantages to real-time and batch processing?

Chapter 7   Part Numbering and Coding

7–1. What are the primary objectives of a part numbering system for manufacturing?

7–2. What are the benefits to significant part numbering systems?

7–3. What are the disadvantages to significant systems?

7–4. Describe the six groups of part identification systems? What are the pros and cons of each?

7–5. Discuss the reasons for the high cost of part number reporting errors.

7–6. Explain the ways material reporting errors can be reduced.

7–7. Describe the fundamental problem with nonsignificant part numbers.

7–8. What is a group technology code?

7–9. Discuss the reasons for using polycodes, monocodes, and hybrids in a group technology project.

7–10. Why are bar codes being so widely used?

7–11. Why do pharmaceutical companies use the CODABAR barcode system?

Chapter 8   Bills of Material
and the Manufacturing Link

8–1. Describe an indented bill of materials.

8–2. Where is a batch bill most commonly used?

8–3. Describe a summarized bill of materials.

8–4. Visit a company or use your own and determine the types of bills used and who uses each type.

8–5. What are the three major subdivisions of bills?

8–6. Describe a planning bill and why it is required for material planning.

8–7. Why are Add/Delete bills a problem for manufacturing planning?

8-8. Why has MRP changed material planning to make the bill of materials so important?

8-9. Which engineering changes are hardest to make? Why?

8-10. What can engineering do to make changes easier to implement?

8-11. Describe the reasons for issuing a new part number rather than simply changing an existing part.

8-12. What are some of the advantages of making block changes?

Chapter 9   Project Planning and Control

9-1. List the four phases of a project and briefly describe each.

9-2. What should the project definition encompass?

9-3. Why is an information flow diagram important in project planning?

9-4. Describe the problems with Gantt or bar charts for project planning.

9-5. Discuss the advantages of the network diagram or bubble chart.

9-6. Which project scheduling technique is most effective for research projects? Why?

9-7. Can Critical Path Method (CPM) be used for new product development projects? Why?

9-8. Describe the advantages of using Project Resource Requirements Planning.

9-9. Discuss the three activity crashing philosophies.

9-10. How can PRRP assist in project budgeting?

Chapter 10   Lead Time and Setup Time

10-1. Compute the Economic Order Quantity given the following:

Fixed ordering cost	$50.00
Inventory carrying cost	32%
Total annual demand	$30,000

10-2. If the ordering cost in Problem 10-1 is 80 percent setup cost and the setup takes 2.6 hours, what is the EOQ reduction from reducing the setup to 15.6 minutes?

10-3. What setup time is required to make a production lot of one (1) economical?

10-4. Explain why reductions in setup time result in lead-time reductions.

10-5. Describe the lead-time whirlpool. Why does it suck in the unwary?

10-6. Discuss the methods or tactics needed to keep the lead-time whirlpool out of a manufacturing plant.

10-7. Discuss the differences between load and capacity in terms of the work in a manufacturing work center.

10-8. What effect do setup times have on capacity? Describe the situation in detail.

10-9. Describe the process required to cut lead times in a make-to-stock environment by one half. Will this work in a make-to-order environment? Why?

10-10. What would you recommend to a company who makes 80 percent of the end product, whose factory is working six days per week, three shifts per day, and has a proposal in increased safety stock by $250,000 to improve customer service from 70 to 95 percent? Will this work?

Chapter 11   Manufacturing Engineering

11-1. Discuss the role of manufacturing engineering in modularization of the product.

11-2. What constitutes current conditions for a methods improvement study?

11-3. Explain the purpose of an operations analysis.

11-4. Discuss the difference between material handling and storage analysis.

11-5. Describe the role of routings in a manufacturing control system.

11-6. Explain the process used to develop a GT coding system for parts.

11-7. Compare a plant layout using common routings and flows (cells) to a discrete or conventional plant layout for:
Investment in machinery and labor
Lead times
Customer service
Flexibility
Productivity

11-8. Describe the real objective of layout optimization.

11-9. What factors does a manufacturing engineer use to determine what to put into cells and what to leave discrete?

11-10. Describe the concept of short-horizon scheduling.

11-11. Describe the Kanban system and how it works.

11-12. What is the effect of large lot sizes on a Kanban system? Discuss.

Chapter 12   Plant Engineering

12-1. Describe the importance of plant layouts to:
Manufacturing engineering
Plant engineering

12-2. What is the effect on lead times of an effective preventive maintenance system?

12-3. Discuss the costs associated with breakdowns.

12-4. Determine the percentage of failures from Figures 12-1, 12-2, and 12-3 you feel can be controlled by an effective preventive maintenance program. Support that position.

12-5. Discuss the four elements of an effect maintenance program.

12-6. Visit a plant and evaluate the effectiveness of the maintenance program. What do you suggest to improve it? What are they doing well?

12-7. What is the effect on a 1200kw rated transformer loaded with 1100kw resistive load and 82 percent power factor? What would you do about it?

12-8. What would you set as the contract capacity for electricity for a manufacturing plant whose connected load is 3000kw but which has a normal average running draw of 1200kw with a 22 percent maximum swing? Why? What would a 10 percent annual growth do to your answer?

12-9. What is the purpose of doing an energy balance on a plant? Discuss that position.

12-10. What is the effect of doubling the insulation in the roof of a plant with an Rll roof? Discuss the effects of roof temperature, plant energy usage, and ventilation.

Chapter 13   Forecasting and Machine Uptime

13-1. Under what conditions is a historical forecast reliable?

13-2. Discuss the reasons that forecasts are more accurate for large groups and over short horizons.

13-3. Visit a company or use your own and list the source, number, and types of forecasts prepared.

13-4. What is the trend formula for the sales data below for a single product item?

Year	1	69,600
	2	99,400
	3	133,000
	4	151,200
	5	175,000

13-5. Using the correlation coefficient, compute the range of and the forecast for year 6.

13-6. What forecasting method would you recommend to compute an estimate of the mean time between failures? Why?

13-7. Given the data:

Total running hrs	Failure
623	1
980	2
1439	3
1753	4

Forecast the next expected failure and compute the expected accuracy of the forecast.

13-8. Using the data in Problem 13-7, compute the MTBF using the Kalman filter technique. Discuss the accuracy of your answer.

13-9. Under what conditions is the Weibull function appropriate for computing the life of an item? Would it be appropriate for:

Gear life

Relay life

Press life

Process controller life

Discuss your position on each.

13-10. What is the major cause of machine failure? Why?

13-11. Visit a nearby manufacturing company and determine what they find to be the major causes of breakdowns. Discuss what to do about each.

13–12. Discuss the effects of failure analysis on the forecasting techniques used to predict failures.

Chapter 14   Long-Range Facilities Planning

14–1. Visit a growing manufacturing company and determine the growth pattern in square feet. Document that growth and develop a forecast based on history.

14–2. Obtain capacity data for all work centers in the plant visited in Problem 14–1. How could capacity be forecast for the company? Discuss.

14–3. Develop a forecast of space needs based upon forecast capacity.

14–4. Discuss the comparison between the ABC company data and the company visited. Do they need like methods or different ones?

14–5. Given the growth pattern of the company visited, develop a plan to expand the facility based upon capacity considerations.

14–6. What is the space efficiency correction factor needed in the plant visited?

14–7. Block plan the expansion and discuss the various alternative layouts.

14–8. What would you do to reduce energy consumption in this expansion? Why?

14–9. Develop a cost estimate for the expansion forecast. Discuss the likely estimate error range.

14–10. Discuss the improvements in forecasting space using the business plan as a base.

Chapter 15   The Just-in-Time Philosophy

15–1. Discuss just-in-time as it applies to a single manufacturing work center. Include work-in-process, lead time, lot size, and setup times.

15–2. Many companies using JIT view engineering as direct labor. Explain this view.

15–3. Visit a local manufacturing company and determine how long it takes in each department to process an order. Include waiting or queue time. Using these data, determine the maximum inventory turns achievable in that environment.

15–4. Compare the maximum turns to the actual average for this company. Discuss the difference between these values.

15–5. What is the average cost of carrying the inventory and the potential savings achievable from a JIT program in the company visited?

15–6. Discuss the relationship between JIT and the MRP II system. Is there a way for each to help the other?

15–7. What effect does quality have on a JIT program and on the manufacturing control system? How do MRP II and JIT treat quality problems?

15–8. Discuss JIT as a strategy given two conditions:
   a. With full management support and understanding
   b. Where top management thinks JIT is a purchasing program

15–9. Describe the role of engineering in a JIT program.

15–10. Compute the potential benefits to engineering alone from effective integration with manufacturing.

BIBLIOGRAPHY

Chapter 1 Engineering in a Manufacturing Company

Alford, L. P., and J. R. Bangs, eds. *Production Handbook*. The Ronald Press, New York, 1953.

Begeman, Myron L. *Manufacturing Processes*. John Wiley & Sons, New York, 1942.

Dossett, Lawrence S. "Engineering Joins the MRP Crusade," APICS Conference Proceedings, 1976.

Drucker, Peter F. *Management: Tasks—Responsibilities—Practices*. Harper & Row, New York, 1973.

Kemper, John Dustin. *The Engineer and his Profession*. Holt, Rinehart and Winston, New York, 1967.

Merrick, Charles M., ed. *ASME Management Division History*. American Society of Mechanical Engineers, New York, 1984.

Peters, Thomas J., and Robert H. Waterman, Jr. *In Search of Excellence*. Harper & Row, New York, 1982.

Strassman, Paul W. *Risk and Technological Innovation*. Cornell University Press, Ithaca, NY, 1959.

Chapter 2 The Manufacturing Control System

Mather, Hal F., and George W. Plossl. *The Master Production Schedule*, 2nd ed. George Plossl Educational Services, Atlanta, GA, 1977.

Orlicky, Joseph A. *Material Requirements Planning*. McGraw-Hill, New York, 1975.

Orlicky, Joseph A. "Closing the Loop With Pegged Requirements and the Firm Planned Order," *Production and Inventory Management*, APICS, First Quarter, 1975.

Plossl, George W. "How Much Inventory is Enough?" *Production and Inventory Management*, APICS, Second Quarter, 1971.

Plossl, George W. *Manufacturing Control: The Last Frontier for Profits*. Reston Publishing Co., Reston, VA, 1973.

Plossl, George W. *Production and Inventory Control: Applications*. George Plossl Educational Services, Inc., Atlanta, GA, 1983.

Plossl, George W. *Production and Inventory Control: Principles and Techniques*. Prentice-Hall, Inc., Englewood Cliffs, NJ, 1985.

Plossl, George W., and W. Evert Welch. *The Role of Top Management in the Control of Inventory*. Reston Publishing Co., Reston, VA, 1979.

Scheele, Evan D., William L. Westerman, and Robert J. Wimmert. *Principles and Design of Production Control Systems*. Prentice-Hall, Englewood Cliffs, NJ, 1960.

Wight, Oliver W. *Production and Inventory Management in the Computer Age*. Cahners Publishing Company, Boston, 1974.

Chapter 3   Design Engineering

Alford, L. P., and J. R. Bangs, eds. *Production Handbook*. The Ronald Press, New York, 1953.

Getzels, J. W., and Philip W. Jackson. *Creativity and Intelligence*. John Wiley & Sons, New York, 1962.

Kemper, John Dustin. *The Engineer and His Profession*. Holt, Rinehart and Winston, New York, 1967.

Jewkes, J., D. Sawers, and R. Stillerman. *The Sources of Invention*. Macmillan, New York, 1958.

Marting, E., ed. *Developing a Product Strategy*. American Management Association, New York, 1959.

Pascale, Richard Tanner, and Anthony G. Athos. *The Art of Japanese Management*. Warner Books, New York, 1981.

Plossl, George W. *Production and Inventory Control: Principles and Techniques*. Prentice-Hall, Inc., Englewood Cliffs, NJ, 1985.

Whyte, William H. *The Organization Man*. Doubleday & Company, Garden City, New York, 1956.

Chapter 4   Computer Systems

Chestnut, Harold. *Systems Engineering Tools*. John Wiley & Sons, New York, 1965.

Kernighan, Brian W., and Dennis M. Ritchie. *The C Programming Language*. Prentice-Hall, Inc., Englewood Cliffs, NJ, 1978.

Leventhal, Lance A. *Z80 Assembly Language Programming*. Osborn/McGraw-Hill, Berkeley, CA, 1979.

Osborne, Adam. *An Introduction to Microcomputers*, Vol. I, *Basic Concepts*. Osborne/McGraw-Hill, Berkeley, CA, 1980.

Ralston, Anthony, ed. *Encyclopedia of Computer Science*. Van Nostrand Reinhold, New York, 1976.

Sage, Andrew P., ed. *Systems Engineering: Methodology and Applications*. IEEE Press, New York, 1977.

Silver, Edward A. "The Use of Programmable Calculators in Inventory Management," *Production and Inventory Management*, APICS, Fourth Quarter, 1979.

Chapter 5   Computer Programming

Harris, L. Dale. *Numerical Methods Using Fortran*. Charles E. Merrill, Columbus, Ohio, 1964.

Kernighan, Brian W., and Dennis M. Ritchie. *The C Programming Language*. Prentice-Hall, Inc., Englewood Cliffs, NJ, 1978.

Jensen, Kathlene, and Niklaus Wirth. *Pascal User Manual and Report*. Springer-Verlag, New York, 1974.

Leventhal, Lance A., *Z80 Assembly Language Programming*. Osborn/McGraw-Hill, Berkeley, CA, 1979.

Richardson, Gary L., Charles W. Butler, and John D. Tomlinson. *A Primer on Structured Program Design*. Petrocelli Books, New York, 1980.

Scanlon, Leo J. *8086/88 Assembly Language Programming*. Robert J. Brady Co., Bowie, MD, 1984.

Yourdon, Edward. *Managing the System Life Cycle*. Yourdon Press, Inc., New York, 1982.

Chapter 6   Computer System Interfaces

Altos ACS8000 Hardware Operating Manual, Version 1.10. Altos Computer Systems, San Jose, CA, 1982.

Birchfield, E. B., and H. H. King. "Product Definition Data Interface," Synergy '84 Conference Proceedings, Chicago, APICS, 1984.

Foundyller, Charles M. *Turnkey CAD/CAM Computer Graphics*. Daratech Associates, Cambridge, MA, 1981.

Hanifen, Leo E. "Taking the Slash Out of CAD/CAM: True Design Manufacturing Integration," Autofact 5 Conference Proceedings, Chicago, SME, 1983.

IBM Synchronous Data Link Control, General Information GA27-3093-1. IBM Corporation, White Plains, New York.

IBM 3650 Retail Store System Loop Interface, OEM Information GA27-3098-0. IBM Corporation, White Plains, New York.

Chapter 7   Part Numbering and Coding

Arn, E. *Group Technology*. Springer-Verlag, Heidelberg, 1975.

El Gomayel, J., and R. Phillips. *Group Technology Applied to Product Design*. Purdue University Press, West Lafayette, Indiana, 1977.

Hyer, Nancy Lea, ed. *Group Technology at Work*. Society of Manufacturing Engineers, Dearborn, Michigan, 1984.

Levulis, R. *Group Technology—A* Review of the State of the Art in the United States. K. W. Tunnell Co., Chicago, IL, 1978.

Plossl, George W. *Manufacturing Control: The Last Frontier for Profits*. Reston Publishing Co., Reston, VA, 1973.

Plossl, George W. *Production and Inventory Control: Applications*. George Plossl Educational Services, Inc., Atlanta, GA, 1983.

Plossl, George W. *Production and Inventory Control: Principles and Techniques*. Prentice-Hall, Inc., Englewood Cliffs, NJ, 1985.

Chapter 8   Bills of Material and the Manufacturing Link

Bourke, Richard W. *Bill of Materials: The Key Building Block*. Bourke and Associates, Inc., Pasadena, CA, 1975.

Johnson, G. A. "Reviewing Part Coding and Classification," *Production and Inventory Management Review*, December 1982.

Mather, Hal F. "A Bill of Materials for All Seasons," *News Note #41*, George Plossl Educational Services, Inc., July 1982.

Mather, Hal. *Bills of Materials: Recipes and Formulations*. Wright Publishing Co., Atlanta, GA, 1982.

Orlicky, Joseph A., George W. Plossl, and Oliver W. Wight. "Structuring the Bill of Material for MRP," *Production and Inventory Management*, APICS, 4th Quarter, 1972.

Plossl, George W. *Manufacturing Control: The Last Frontier for Profits*. Reston Publishing Co., Reston, VA, 1973.

Plossl, George W. *Production and Inventory Control: Applications*. George Plossl Educational Services, Inc., Atlanta, GA, 1983.

Plossl, George W. *Production and Inventory Control: Principles and Techniques*. Prentice-Hall, Inc., Englewood Cliffs, NJ, 1985.

Chapter 9   Project Planning and Control

Aquilano, Nicholas J., and Dwight E. Smith. "A Formal Set of Algorithms for Project Scheduling with Critical Path Scheduling—Materials Requirements Planning," *Journal of Operations Management*, November 1980.

Battersby, A. *Network Analysis for Planning and Scheduling*. St. Martins Press, New York, 1964.

Chase, R. B., and N. J. Aquilano. *Production and Operations Management.* Richard B. Irwin, Homewood, IL, 1981.

Davis, E. W., ed. *Project Management: Techniques, Applications and Managerial Issues.* Industrial Engineering and Management Press, Norcross, GA, 1983.

Kerzner, Harold. *Project Management, A Systems Approach to Planning, Scheduling and Controlling.* Van Nostrand Reinhold Co., New York, 1979.

Moder, Joseph J., Cecil R. Phillips, and Edward W. Davis. *Project Management with CPM, PERT and Precedence Diagramming.* Van Nostrand Reinhold Company, New York, 1983.

Weist, Jerome D., and Ferdinand K. Levy. *A Management Guide to PERT/CPM.* Prentice-Hall, Inc., Englewood Cliffs, NJ, 1969.

Chapter 10 Lead Time and Setup Time

Axsater, S. "Economic Order Quantities and Variations in Production Load," *International Journal of Production Research*, May–June 1980.

Banks, Jerry, and C. L. Hohenstein. "Simplification of the Economic Order Quantity Equation," *Journal of Purchasing and Materials Management*, Summer 1981.

Harty, J. D., G. W. Plossl, and O. W. Wight. "Management of Lot-size Inventories," APICS Special Report, September 1963.

Plossl, George W., "The Semiconductor Story," *Newsletter #20*, George Plossl Educational Services, Atlanta, GA, 1976.

Plossl, George W., *Production and Inventory Control: Applications.* George Plossl Educational Services, Inc., Atlanta, GA, 1983.

Plossl, George W. "The Semiconductor Story—A Sequel," *News Note 49*, George Plossl Educational Services, Atlanta, GA, 1983.

Plossl, George W. *Production and Inventory Control: Principles and Techniques.* Prentice-Hall, Inc., Englewood Cliffs, NJ, 1985.

Chapter 11 Manufacturing Engineering

Alford, L. P., and J. R. Bangs, eds. *Production Handbook.* The Ronald Press, New York, 1953.

Dickie, H. F. "ABC Inventory Analysis Shoots for Dollars," *Factory Management and Maintenance*, July 1951.

Plossl, George W. *Production and Inventory Control: Applications.* George Plossl Educational Services, Inc., Atlanta, GA, 1983.

Plossl, George W. *Production and Inventory Control: Principles and Techniques.* Prentice-Hall, Inc., Englewood Cliffs, NJ, 1985.

Schonberger, Richard J. *Japanese Manufacturing Techniques.* The Free Press, New York, 1982.

Shingo, Shigeo. *Study of Toyota Production System.* Japan Management Association, Tokyo, 1981.

Chapter 12   Plant Engineering

Brownell, Donald R., and Mark S. Miller. "Model-based Service Parts Forecasting," APICS Conference Proceedings, New Orleans, La. 1983.

"Plant Maintenance Management System," IBM Publication No. E20-0124-0. IBM Corporation, White Plains, New York.

Plenert, G. J. "Bill of Energy," APICS Conference Proceedings, Chicago, Il., 1982.

Plossl, George W. *Production and Inventory Control: Applications.* George Plossl Educational Services, Inc., Atlanta, GA, 1983.

Plossl, George W. *Production and Inventory Control: Principles and Techniques.* Prentice-Hall, Inc., Englewood Cliffs, NJ, 1985.

Sack, Thomas F. *A Complete Guide to Building and Plant Maintenance*, 2nd ed. Prentice Hall, Inc., Englewood Cliffs, NJ, 1971.

Swanson, Jim. "Machine Tools That Keep Working," *Production Engineering*, Vol. 27, NO. 8, November 1980.

Chapter 13   Forecasting and Machine Uptime

Brown, Robert G. *Statistical Forecasting for Inventory Control.* McGraw-Hill, New York, 1959.

Brownell, Donald R., and Mark S. Miller. "Model-based Service Parts Forecasting," APICS Conference Proceedings, New Orleans, La., 1983.

Flowers, A. Dale. "A Simulation Study of Smoothing Constant Limits for an Adaptive Forecasting System," *Journal of Operations Management*, Vol. 1, No. 2, November 1980.

Lobdill, Jerry. "Kalman Mileage Predictor-Monitor," *Byte*, Vol. 6, No. 7, July 1981.

Kalman, R. E. "A New Approach to Linear Filtering and Prediction Problems," *Journal of Basic Engineering*, March 1960.

Gelb, Arthur, et al. *Applied Optimal Estimation.* MIT Press, Cambridge, MA, 1974.

Plossl, George W. "Getting the Most from Forecasts," *Production and Inventory Management*, APICS, First Quarter, 1973.

Plossl, George W. *Production and Inventory Control: Applications.* George Plossl Educational Services, Inc., Atlanta, GA, 1983.

Plossl, George W. *Production and Inventory Control: Principles and Techniques.* Prentice-Hall, Inc., Englewood Cliffs, NJ, 1985.

Weibull, W. "A Statistical Representation of Fatigue Failures in Solids," *Transactions of the Royal Institute of Technology*, No. 27, Stockholm, 1949.

Weibull, W. "A Statistical Distribution Function of Wide Applicability," *Applied Mechanics*, September 1951.

Chapter 14 Long-Range Facilities Planning

Plossl, George W. *Production and Inventory Control: Applications.* George Plossl Educational Services, Inc., Atlanta, GA, 1983.

Plossl, George W. *Production and Inventory Control: Principles and Techniques.* Prentice-Hall, Inc., Englewood Cliffs, NJ, 1985.

Sack, Thomas F. *A Complete Guide to Building and Plant Maintenance*, 2nd ed. Prentice Hall, Inc., Englewood Cliffs, NJ, 1971.

Sule, Dileep R., "Simple Methods for Uncapacitated Facility Location/Allocation Problems," *Journal of Operations Management*, Vol. 1, No. 4, May 1981.

Chapter 15 The Just-in-Time Philosophy

APICS. "Zero Inventories Crusade," American Production and Inventory Control Society, Falls Church, VA, 1984.

Burnham, John M. *Japanese Productivity: A Study Mission Report.* American Production and Inventory Control Society, 1983.

Hall, Robert W. "Driving the Productivity Machine: Production Planning and Control in Japan," American Production and Inventory Control Society, Falls Church, VA, 1981.

Nakane, Jinichiro, and Robert W. Hall. "Transferring Production Control Methods Between Japan and the United States," APICS International Conference Proceedings, Boston, MA, 1981.

Plossl, George W. "Lessons in Japanese—Part 1," *News Note #40*, George Plossl Educational Services, Inc., Atlanta, GA, May 1982.

Plossl, George W. *Production and Inventory Control: Applications.* George Plossl Educational Services, Inc., Atlanta, GA, 1983.

Plossl, George W. *Production and Inventory Control: Principles and Techniques.* Prentice-Hall, Inc., Englewood Cliffs, NJ, 1985.

Plossl, Keith R. "Lessons in Japanese—Part 3," *News Note #43*, George Plossl Educational Services, Inc., Atlanta, GA, November 1982.

Plossl, Marion L. "Lessons in Japanese—Part 2," *News Note #42*, George Plossl Educational Services, Inc., Atlanta, GA, September 1982.

Schonberger, Richard J. *Japanese Manufacturing Techniques.* The Free Press, New York, 1982.

Shingo, Shigeo. *Study of Toyota Production System.* Japan Management Association, Tokyo, 1981.

Wight, Oliver W. *MRP II—Unlocking America's Productivity Potential.* The Book Press, Brattleboro, VT, 1981.

INDEX